VINCENNES UNIVERSITY LIBRARY

Mutation

THE HISTORY OF AN IDEA
FROM DARWIN TO GENOMICS

ALSO FROM COLD SPRING HARBOR LABORATORY PRESS

Davenport's Dream: 21st Century Reflections on Heredity and Eugenics edited by Jan A. Witkowski and John R. Inglis

Evolution by Nicholas H. Barton, Derek E.G. Briggs, Jonathan A. Eisen, David B. Goldstein, and Nipam H. Patel

Francis Crick: Hunter of Life's Secrets by Robert Olby

Max Perutz and the Secret of Life by Georgina Ferry

Mendel's Legacy: The Origin of Classical Genetics by Elof Axel Carlson

Times of Triumph, Times of Doubt: Science and the Battle for Public Trust by Elof Axel Carlson

What a Time I Am Having: Selected Letters of Max Perutz edited by Vivien Perutz

Mutation

THE HISTORY OF AN IDEA
FROM DARWIN TO GENOMICS

Elof Axel Carlson

COLD SPRING HARBOR LABORATORY PRESS
Cold Spring Harbor, New York • www.cshlpress.com

Mutation
THE HISTORY OF AN IDEA FROM DARWIN TO GENOMICS

Printed in the United States of America

Publisher	John Inglis
Acquisition Editor	Judy Cuddihy
Director of Development, Marketing, & Sales	Jan Argentine
Developmental Editor	Judy Cuddihy
Project Manager	Inez Sialiano
Permissions Coordinator	Carol Brown
Production Editor	Kathleen Bubbeo
Production Manager	Denise Weiss
Marketing & Sales Manager	Elizabeth Powers
Cover Designer	Ed Atkeson

Front cover artwork: The rediscovery in 1900 of Gregor Mendel's mid-19th century breeding experiments with pea plants initiated a major shift in thinking about the nature of mutation from Darwin's fluctuating variations to a discontinuous model of evolution by mutation. Cover image source: VectorStock/Mombeka.

Library of Congress Cataloging-in-Publication Data

Carlson, Elof Axel.
Mutation : the history of an idea from Darwin to genomics / Elof Axel Carlson.
p. cm.
Includes bibliographical references and index.
ISBN 978-1-936113-30-9 (hard cover : alk. paper)
1. Mutation (Biology) 2. Mutation (Biology)–Research. 3. Genomics. I. Title.

QH460.C37 2011
576.5′49–dc22

2011007489

10 9 8 7 6 5 4 3 2 1

All World Wide Web addresses are accurate to the best of our knowledge at the time of printing.

All Cold Spring Harbor Laboratory Press publications may be ordered directly from Cold Spring Harbor Laboratory Press, 500 Sunnyside Blvd., Woodbury, New York 11797-2924. Phone: 1-800-843-4388 in Continental U.S. and Canada. All other locations: (516) 422-4100. FAX: (516) 422-4097. E-mail: cshpress@cshl.edu. For a complete catalog of all Cold Spring Harbor Laboratory Press publications, visit our website at http://www.cshlpress.com/.

Contents

Preface

IN MY FIRST BOOK ON THE HISTORY of genetics, *The Gene: A Critical History*, I showed how the concept of the gene emerged from a series of contending views and how each of about one dozen disputes was resolved with the preservation of one view and the disappearance of the other. This, my fourth book on historical issues, once again traces the development of a concept, this time "mutation," a term that has undergone significant change in the hands of professional scientists and also become significant in popular culture.

The idea of mutation is rooted in our awareness of change over time. In the life sciences, consideration of change is essential to evolutionary biology and also, perhaps less obviously, to the study of genetics. Ideas or concepts also change, or evolve, over time. Each generation of scholars is bound by the terminology and ideas current in its period but as new tools and approaches generate new findings, the scholars' vocabulary changes to accommodate what has been discovered. This book examines the meaning of mutation when the term was first adopted, how the meaning changed, and why that alteration was forced on the terminology. Among professional scientists, there is further complexity, with the coexistence of contending terms for the same phenomenon used in different fields (e.g., microbial and *Drosophila* genetics). I also show that a somewhat different sense of the term mutation prevails among the general public.

Many scientists tend to be unaware of how their colleagues of many generations ago conceived their field. Examination of this process is the task of an historian but has the added benefit of informing us about the way ideas help or hinder the development of a field of science.

I am grateful to Jan Witkowski for organizing a conference on mutation in May 2010 at the Banbury Center of Cold Spring Harbor Laboratory. It inspired me to think about my own contribution to that conference and to reflect on the history of the concept of mutation. This book is a result of that reflection and research.

I thank the staff of Cold Spring Harbor Laboratory Press for the outstanding work they have done in making this book possible: John Inglis, Publisher; Judy Cuddihy, Acquisition and Developmental Editor; Jan Argentine, Director of Development, Marketing, and Sales; Inez Sialiano, Project Manager; Carol Brown, Permissions Coordinator; Kathleen Bubbeo, Production Editor; Denise Weiss, Production Manager; and Elizabeth Powers, Marketing and Sales Manager. I am especially grateful to Judy Cuddihy and Kathleen Bubbeo for their useful suggestions as my manuscript shifted from a working manuscript to final page proofs. I also thank Krishna Dronamraju, Michael Lynch, and Evelyn Witkin for their helpful discussions about mutation. The Biology Library at Jordan Hall and the Wells Library, both at Indiana University in Bloomington, had the references I needed and provided the comfort to take notes.

ELOF AXEL CARLSON

Mutation

THE HISTORY OF AN IDEA
FROM DARWIN TO GENOMICS

1

A Brief Overview of the Concept of Mutation

TODAY MOST GENETICISTS THINK OF THE TERM "mutation" as a technical term for some change in the individual gene. This concept dates to the early 1920s, and Hermann Joseph Muller (1890–1967) forcefully argued this restricted definition.[1] In this history, I will provide the background to that conclusion and show how this view permeated biological thinking and how it became modified as the attributes of genes became augmented. Breeding analysis was supplemented by cytology, population genetics, physiological attributes of gene expression, developmental studies of the origin of mutations in the life cycle, biochemical studies of the genetic control of biochemical pathways, molecular analysis of gene products, the identification of genes as nucleic acids, the various ways chemical and physical agents act as mutagens, and the analysis of how genes function at the molecular level.

As the attributes and process of mutation were worked out, the potentials for using mutations increased. Mutations entered the thinking of population genetics, evolutionary studies, horticulture, plant and animal commercial development, and human genetics. They also entered politics in the form of worldwide eugenics movements, racist ideology, and genocidal programs, as well as the state control of what could be taught about mutation as it applied to human benefit and social needs. With the discovery of the induction of mutations by X rays in 1927 and the founding of the field of radiation genetics, mutation took on a new significance as the use of radiation in society increased. After the use of atomic bombs in Japan during World War II, the public became conscious of radiation protection and radiation hazards.[2]

In the last decades of the 20th century, molecular interest in mutations led to the Human Genome Project and an attempt to identify all human

Muller, H. J. 1922. Variation due to change in the individual genes. *The American Naturalist*, 56: 32-50.

VARIATION DUE TO CHANGE IN THE INDIVIDUAL GENE[1]

DR. H. J. MULLER

The University of Indiana

I. THE RELATION BETWEEN THE GENES AND THE CHARACTERS OF THE ORGANISM

The present paper will be concerned rather with problems, and the possible means of attacking them, than with the details of cases and data. The opening up of these new problems is due to the fundamental contribution which genetics has made to cell physiology within the last decade. This contribution, which has so far scarcely been assimilated by the general physiologists themselves, consists in the demonstration that, besides the ordinary proteins, carbohydrates, lipoids, and extractives, of their several types, there are present within the cell thousands of distinct substances — the "genes"; these genes exist as ultramicroscopic particles; their influences nevertheless permeate the entire cell, and they play a fundamental role in determining the nature of all cell substances, cell structures, and cell activities. Through these cell effects, in turn, the genes affect the entire organism.

It is not mere guesswork to say that the genes are ultramicroscopic bodies. For the work on *Drosophila* has not only proved that the genes are in the chromosomes, in definite positions, but it has shown that there must be hundreds of such genes within each of the larger chromosomes, although the length of these chromosomes is not over a few microns. If, then, we divide the size of

1 In symposium on "The Origin of Variations" at the thirty-ninth annual meeting of the American Society of Naturalists, Toronto, December 29, 1921.

H.J. Muller limits the term mutation. In this article Muller asks geneticists to consider the term mutation as limited to changes within the individual gene. He considered all other phenomena such as polyploidy, aneuploidy, and chromosomal rearrangements to be distinct phenomena. All played a role in evolution, but Muller believed the major events in evolution to be the mutations in individual genes. (Reprinted from Muller HJ. 1922. Variation due to change in the individual gene. Am Nat 56: *32–50.)*

genes and later a comparative genomics of different phyla or organisms of social or evolutionary importance. That process has occupied the first decade of the 21st century. Molecular tools permit geneticists to shift altered genes from species to species; organisms can be engineered to perform specific functions depending on the introduced genes used in the design. Craig Venter (1946–) and colleagues even devised means to replace the genome of one bacterial species with another that was synthesized from a computer printout and annotated with words that Venter called "watermarks" that could be decoded.[3] Few geneticists today would be surprised if a paper appeared announcing the synthesis of prokaryotic cells containing a bare minimum of genes necessary for life to continue. Synthetic nuclei may some day be used to resurrect extinct species or putative common ancestors of related species.

In this book, Chapters 2–4 discuss nonmolecular or classical genetic approaches to mutation. Chapters 5–8 cover biochemical and molecular approaches to mutation. Chapter 9 covers the emerging field of genetic engineering and comparative genomics, and Chapter 10 looks at the social uses and abuses of mutation. Finally, Chapter 11 explores the implications that mutation has for the philosophy and history of science.

Mutation Evolved as a Concept

Mutation, of course, involves change; but our understanding of that change is influenced by the time we live in. To Darwin, most changes of evolutionary significance were "fluctuating variations," mostly subtle, providing the raw material on which natural selection acted. What geneticists think of as mutations today are closer to what Francis Galton (1822–1911) thought them to be in the late 19th century. For him, mutations were shifts in dramatic, qualitative, or discontinuous expression, like albinism from pigmented populations or the sudden appearance of what breeders called "sports." For example, Ancon sheep—a short-legged line with crooked forelegs—were frequently cited as having a pedigreed origin from a spontaneous appearance of a single short-legged lamb.

Complicating this distinction between fluctuating variations and sports was the distinction between the terms "variation" and "heredity." Variations were seen as the changes from the normal state of a trait, and they could be either more intense or less so in expression. Heredity was seen as the sum of species traits. Heredity and variation were considered to be separate phenomena. It should be kept in mind that in the last quarter of the 19th century when these ideas were widely circulated, Mendelism was either forgotten or not considered relevant to the problem of mutation by those scientists discussing it.

What forced a change in the thinking about mutation was the rediscovery of Mendelism in 1900 and the insistence by William Bateson (1861–1926) in 1894 that variations could be discontinuous and might be associated with the origin of species. This led to a bitter dispute between the loyalists to Darwinian fluctuating variations and Bateson's school of Mendelians, who saw "discontinuous variations" of a more dramatic nature as significant for evolution. The first battles were largely fought in Great Britain, with W.F.R. Weldon (1860–1906) and Karl Pearson (1857–1936) supporting the Darwinian fluctuation model and Bateson supporting the discontinuous model of evolution by mutation. Francis Galton supported both models and contributed articles championing both, sometimes to the annoyance or surprise of those who thought he was solidly behind one of these approaches.[4]

The Mendelian school emerged from the work of Hugo de Vries (1848–1935), Carl Correns (1864–1933), and Erich von Tschermak-Seysenegg (1871–1962) in 1900. Their independent rediscoveries of Mendel's laws of transmission of inherited traits focused an interest on discontinuous rather than blending or barely perceptible quantitative changes in expression of character traits. The initial traits used were mostly preexisting variations of plants known to horticulturists, but de Vries took an interest in the sudden appearance of new forms, including what he thought were "mutations to new species" in the evening primrose, *Oenothera lamarckiana*. This emphasis on discontinuous traits made de Vries and Bateson allies in their conflicts with the "biometric group," as they called themselves, championed by Weldon and Pearson.

Whereas these disputes occupied European geneticists from the 1890s to the 1920s, another school emerged that used a different approach to studying mutation. Edmund Beecher Wilson (1856–1939) championed a cytological approach to understanding heredity, particularly through sex determination. Several American biologists, including Clarence McClung (1870–1946), Thomas Montgomery (1873–1912), and Nettie Maria Stevens (1861–1912), looked at an unusual chromosome or chromosome-like body in the nucleus first reported in 1891 in Germany by Hermann Henking (1858–1942). Wilson clarified these contributions by proposing to name this "accessory chromosome" as an X chromosome; in some species, he noted that a smaller Y chromosome accompanied it. Some species of bugs just used XX female and X male, whereas others used XX female and XY male.[5]

Wilson's ideas that sex determination was discontinuous and had a mechanistic basis through sex chromosomes influenced his colleague Thomas Hunt Morgan (1866–1945), who was skeptical of most models associating hypothetical mutations with evolution or other biological traits. Morgan started out seeking new mutations to produce new species because de Vries's approach lent itself to experimental testing. Morgan was rewarded with the discovery of several discontinuous mutations that arose in his fruit fly stocks; however, no new species arose. These mutations were qualitatively different or striking in their departure from the normal expression of a trait. None showed the minor fluctuations that the biometric school favored based on field studies of species. By 1913, Morgan and his students were arguing that new mutations arose spontaneously, that some were associated with a new form of heredity that they called "sex-limited," and that these factors could be mapped linearly on a representation of the X chromosome.[6] The fly group's cytogenetic approach proved far more fruitful in studying mutations than did the more theory-based approaches of the biometric group. Curiously, Bateson felt unconvinced that this cytological approach would be sufficient to explain mutations or the relation of genetics to evolution. His skepticism was based on the relatively small number of chromosomes in a cell and the many thousands of traits in individuals. Instead of genes aligned along a chromosome, Bateson imagined numerous isolated hereditary factors with combining attractions or repulsions like magnetic dipoles in a nuclear sac.

By the 1920s, the concept of mutation had shifted to Morgan's way of thinking. Genes experienced random mutations. These could be followed to their origins in sperm or egg cells or to perifertilization events shortly before or shortly after fertilization. The properties of mutations were becoming clearer. A gene could mutate in different ways, leading to multiple allelic series. Some gene mutations led to a complete loss of function (as in a human albino's loss of pigment), and those were attributed to either losses of the gene by deletion or changes within the gene that caused it to fail to function. Some mutations showed only traces of activity, and these could be measured by dosage studies of genes. Some had substantial losses leading to variations that departed from a normal trait, like red eyes with apricot, eosin, or other tints of the red pigments. By the 1930s, the physiology of eye color mutations was emerging. Two pigments were involved, and one (the formation of brown eye pigment) was biochemically related to the kynurenine pathway. These pigments were deposited in the eyelets (ommatidia) of the compound eye. The white-eyed series of multiple alleles involved this deposition protein lining the ommatidia.

Early Studies Tried to Induce Gene Mutations

Scientists tried to induce mutations in the early 20th century. Albert F. Blakeslee (1874–1954), James W. Mavor (1883–1963), Morgan, and Fernandus Payne (1881–1977) failed to find convincing evidence that X rays or radium induced mutations. Some used alcohol or ether as possible agents to induce mutation. Mutation rates were not measured until 1919 because they were so rare that no one knew what magnitude to use for setting up an experiment. Muller used a quantitative approach to find mutations by using temperature (a 10°C difference for raising flies), but he also retried the X-ray approach after reading the literature on radiation biology. He designed a stock that would detect a class of mutations that killed the males carrying it. He called these "X-linked lethal mutations." His stock design found such X-linked lethals in abundance at the doses he used (today estimated as about 4000 roentgens). Muller began his X-ray experiments in the fall of 1926 and announced his discovery[7] in a paper in *Science* in 1927 (with no detailed data to back up his claim). That evidence, however, was provided in abundance in his 1927 presentation at the Fifth International Congress of Genetics in Berlin.[8] Lewis Stadler (1896–1954), using mutations induced in cereal grains by X rays, confirmed Muller's work. The field of radiation genetics rapidly emerged in the 1930s and 1940s and showed that X rays induced both changes in the individual genes and a much more abundant class of changes associated with chromosome breakage and chromosome rearrangements (deletions, inversions, and translocations), as well as a category of rearrangements leading to aneucentric chromosomes (having no centromere or two centromeres).

These aneucentric chromosome rearrangements played an important part in explaining a category of fertilized eggs that failed to develop ("dominant lethals") in fruit flies. The same mechanism (called the breakage–fusion–bridge cycle by Barbara McClintock [1902–1992]) could be used to explain the symptoms of radiation sickness among those residents exposed to heavy doses of radiation in the atomic bombing of Hiroshima and Nagasaki. (A-bomb radiation induced aneucentric chromosomes in dividing tissues, resulting in blistered skin, ruptured capillaries, anemia, diminished immune response, and ulcerative peritonitis.) The concerns regarding mutations (to posterity) and radiation sickness or the induction of future cancers proved troubling to society, and considerable debate among scientists emerged from 1945 to the 1970s regarding the hazards of fallout from nuclear weapons testing, medical use of radiation, and industrial use of radiation (especially to generate electricity).

Note how the idea of the classical gene, supplemented by cytology, shifted the way in which mutation was seen both by scientists and the public. In its social applications, that classical gene was assigned human behavioral genetic roles with a paucity of experimental evidence. This led to abuses by the eugenics movement, which bolstered its ideology of human betterment through genetic arguments. Feeblemindedness, pauperism, and other social traits quickly became genetic traits, as fixed and easy to conceive as Mendelian mutations in fruit flies.[9] In contrast, the findings of laboratory studies of mutation had substantial experimental work to back up concepts like dosage compensation for X-linked traits; the use of terms like amorphs, hypomorphs, and neomorphs to describe the emergence of new mutations; and the analysis of complex systems of alleles with properties described as pseudoallelism or complementation in addition to classical multiple allelism.

Mutation Is Molecularized

The presentation of the double-helix model of deoxyribonucleic acid (DNA) in 1953 by James D. Watson (1928–) and Francis H.C. Crick (1916–2004) shifted the concept of the gene, in scientific minds, to a sequence of nucleotides. A genetic code was worked out quickly—a triplet nucleotide sequence (codon) of DNA specified a specific amino acid or some functional attribute of the translation process by which reading begins and stops along a length of nucleic acid. Changes in a codon were associated with mutations. Some were simple amino acid or nucleotide replacements. Some represented a shift in reading sets of three nucleotides, leading to large segments of gibberish in the protein product eventually produced by the conversion of information (sequences of DNA nucleotides) to product (sequences of protein amino acids).[10]

In the case of evolution studies, this molecular idea of the gene led to discussions on the role of "neutral mutations" that either involved codon changes that did not lead to amino acid replacements or mutations occurring in introns (regions of the gene not associated with the assembling of the protein product). Also, not all replacements of amino acids in a protein had harmful or beneficial effects on that protein's basic function. Some were neutral. Such changes, however, served as useful markers to measure mutation rates over evolutionary time. The molecular analysis of genes in cognate species also allowed geneticists to classify genes by function and identify when such genes were called into action during the life cycle of an organism.

For physicians, a major bonus of the molecular concept of gene mutation was its utility for prenatal diagnosis and for a more long-range strategy of gene replacement therapy in targeted tissues (or even farther into the future, for embryonic cells).[11]

The molecular tools for using gene mutations are now numerous. Plants can be made resistant to fungal or bacterial disease. They can be modified to resist drought, brackish water, or the shearing effects of wind. They can change the commercial qualities of products, increasing the shelf life of fruits and vegetables. They can change the nutritional value of meats, reducing more harmful fats and introducing fats with either a benign or beneficial contribution to human nutrition. As in the history of classical genetics through the early years of the Cold War, the concept of mutation continues to generate public controversy. Sometimes the debates are not resolved, but become eclipsed by the development of new techniques, new tools, and new interests of society. This brief overview puts into perspective the long history of the term "mutation." The following chapters explore the detailed conceptual changes in the idea of mutation, as well as its role in science and society.

References and Notes

1. Muller HJ. 1922. Variation due to change in the individual gene. *Am Nat* **56:** 32–50.
2. Carlson EA. 1981. *Genes, radiation, and society: The life and work of H.J. Muller.* Cornell University Press, Ithaca, NY. See Chapter 24, "Human mutation and the radiation danger," and Chapter 25, "The fallout controversy" (pp. 336–367).
3. Gibson DG, Glass JI, Lartique C, Noskov VN, Chuang R-Y, Algire MA, Benders GA, Montague MG, Ma L, Moodie MM, et al. 2010. Creation of a bacterial cell controlled by a chemically synthesized genome. *Science* **329:** 52–56.
4. For an overview of Bateson's life, personality, and career, see Cock AG, Forsdycke DR. 2008. *Treasure your exceptions: The life and science of William Bateson.* Springer, New York. Also see Chapter 3, "Galton's disciples," in Schwartz J. 2008. *In pursuit of the gene: From Darwin to DNA.* Harvard University Press, Cambridge, MA.
5. See Chapter 7, "The sex chromosomes," and Chapter 9, "The chromosome theory of heredity," in Carlson EA. 2004. *Mendel's legacy: The origin of classical genetics.* Cold Spring Harbor Laboratory Press, Cold Spring Harbor, NY.
6. Allen G. 1978. *T. H. Morgan—The man and his work.* Princeton University Press, Princeton, NJ.
7. Muller HJ. 1927. Artificial transmutation of the gene. *Science* **66:** 84–87.

8. Muller HJ. 1928. The problem of genic modification. *Z Indukt Abstammungs-Vererbungslehre.* **Suppl 1:** 224–260.
9. Witkowski JA, Inglis JR, eds. 2008. *Davenport's dream: 21st century reflections on heredity and eugenics.* Cold Spring Harbor Laboratory Press, Cold Spring Harbor, NY.
10. Cairns J, Stent GS, Watson JD, eds. 1966. *Phage and the origins of molecular biology.* Cold Spring Harbor Laboratory, Cold Spring Harbor, NY.
11. Cowan RS. 2008. *Heredity and hope: The case for genetic screening.* Harvard University Press, Cambridge, MA.

2

Ideas of Mutation before There Was a Mendelian Basis for Genetics

THE TERM "MUTATION" COMES FROM THE LATIN verb *mutare*, which means "to change." It was rarely used by biologists before the 20th century, and a variety of terms were used instead. William Bateson began using the term mutation in 1894, but it became better known with the publication of Hugo de Vries's book *The Mutation Theory* in 1901.[1]

When Charles Darwin (1809–1882) was a young man, our present-day idea of mutations, as alterations of genes or heredity, did not exist. A variety of other terms and ideas were current. Those who were practical breeders of plants and animals for commercial or hobby interests used terms like "keeping up the breed" when describing attempts to maintain one or more characters in a variety or breed. If the trait had no previous history, as in the sudden appearance of a knotted region of a puppy dog's tail, the breeder would refer to its origin as "accidental,"[2] with some implication or belief that body damage could somehow be transmitted. This idea persisted throughout most of the 19th century until it was laid to rest in the 1880s by the works of August Weismann (1834–1914), who demonstrated that such mutilations or accidents had no effect on the "germplasm," the term he applied to the isolated hereditary material that formed the future eggs and sperm. In the 1830s, Darwin's correspondents used terms like "hybrid," "mongrel," or "mixed breeds" interchangeably.[3]

A highly inbred stock of cattle would be said to have "high blood" and fairly consistent traits from many generations of this practice, but their offspring tended to be of "feeble virility."[4] In 1839, after he returned from the voyage of *The Beagle*, Darwin worked for a few years from a home in London before he moved permanently to Down House in Kent. He wrote to many breeders, sending them questionnaires he prepared on breeding,

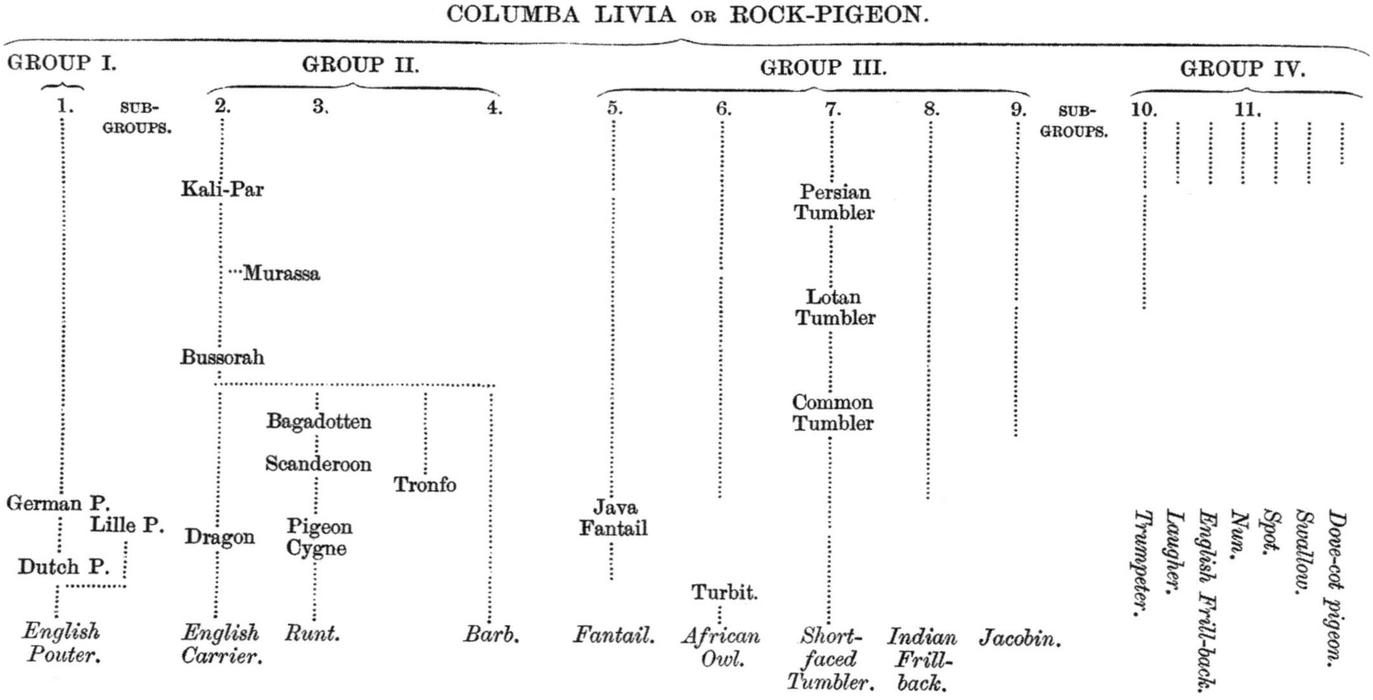

COLUMBA LIVIA OR ROCK-PIGEON.

GROUP I.
1.
SUB-GROUPS.
German P.
Lille P.
Dutch P.
English Pouter.

GROUP II.
2. 3. 4.
Kali-Par
Murassa
Bussorah
Bagadotten
Scanderoon
Tronfo
Dragon
Pigeon Cygne
English Carrier.
Runt.
Barb.

GROUP III.
5. 6. 7. 8. 9.
Persian Tumbler
Lotan Tumbler
Common Tumbler
Java Fantail
Turbit.
Fantail.
African Owl.
Short-faced Tumbler.
Indian Frill-back.
Jacobin.

SUB-GROUPS.

GROUP IV.
10. 11.
Trumpeter.
Laugher.
English Frill-back.
Nun.
Spot.
Swallow.
Dove-cot pigeon.

Charles Darwin offers pigeon breeds as a model for evolution. Darwin identified pigeons as an ideal model of how variations could lead to breeds that all descend from a common origin through conscious selection. The progenitor is the common rock pigeon. All the popular breeds Darwin studied arose by efforts of breeders in different countries over many centuries. (Reprinted from Darwin C. 1868. The variation of plants and animals under domestication. *John Murray, London.)*

variations, and changes they noted in the first or later generations of crosses of plants or animals. At this stage of his career, Darwin did not publicly offer a theory of the origin of variations or regarding his doubts about the fixity of species. He was convinced that species changed and that the demarcation between varieties and species was blurred. He first sketched his idea of those problems in a notebook in 1838; in essays written in 1842 and 1844, he prepared more detailed arguments of his theory of natural selection, but these were not published in his lifetime. His son Francis prepared these essays to be published as part of the 50th anniversary of the publication of *On the Origin of Species* in 1909.[5]

In the 1842 essay, Darwin argued that "an individual organism placed under new conditions sometimes varies in a small degree in very trifling respects such as stature, fatness, sometimes color, health, habits in animals and probably disposition."[6] Darwin was also familiar, from conversations with breeders, with a very different type of variation. These variations were called "sports," and when they arose from a bud upon a plant, they were sometimes referred to as "bud sports." In Darwin's phrasing, "When the individual is multiplied for long periods by buds the variation is yet small, though . . . occasionally a single bud or individual departs widely from its type and continues steadily to propagate by buds such new kind."[7] These two forms of variation—one subtle and barely noticeable to the untrained eye and the other dramatic and readily picked up by amateur breeders—were the source of variations in both domesticated forms and in the wild or nature. But these subtle variations were hard to detect in nature. Darwin noted, "wild animals vary exceedingly little—yet they are known as individuals."[8] Why was there this difference between the hard-to-notice natural variation in species and the confined status of domesticated forms whose variations (among breeders of all types) are quite common? Darwin argued that something about the domesticated environment was causing the more frequently seen variation. In his words, "any and every one of these organisms would vary if the organism were taken away and placed under new conditions."[9] To Darwin, variations had to arise from some sort of mechanism present in the species or variety that reproduced. He rejected the Lamarckian (Jean-Baptiste Lamarck [1744–1829]) model of variations arising in response to the environment in an adaptive way. These variations showed no such relation to their artificial environments. The variations he had seen in the natural habitat among the birds, fish, reptiles, insects, and other animals he examined in his round-the-world trip suggested that something else was going on.

Darwin Introduces a Concept of "Plasticity" Especially in Domesticated Species

In 1844, Darwin used another term to describe this condition—he referred to the organism's "plasticity" under changed environments. Their hereditary nature in response to this plasticity seemed to yield more variations when stressed by domestication, and it "gives way or breaks."[10] The bud sports in plants resembled what Darwin called, in animals, "congenital peculiarities."[11] These were often represented as extra limbs or digits, loss of a tail, or the substitution in fowl of feathers for a coxcomb. They differed in kind to Darwin from the more minor variations that tended to be plus or minus variations of the same trait (e.g., lighter or darker, straighter or more curved, rougher or smoother in texture). They also differed in the suddenness of their abnormality or peculiarity, which came immediately to the attention of a breeder. Such sports were seen in nature, too, but much more rarely, Darwin believed, than among domesticated organisms, a view he shared with breeders whom he contacted at shows or by mail. These differences made sports some sort of pathologies of heredity and not the material on which species could form.

By 1859, when he published his *On the Origin of Species*, Darwin continued his belief that variations under domestication were more frequent than in the wild. He associated one of the causes of this shifted environment under domestication with feeding. Animals eat more when domesticated, and Darwin thought this might be disruptive to the organism's heredity.[12] He claimed that no domesticated form ceases to be variable, even when highly inbred. Darwin distinguished the capacity to be tamed from the capacity to be domesticated. From his readings and discussions with many breeders, zookeepers, and naturalists, he concluded that almost any animal can be tamed but only a small number of animal types (in 1859 about 24) could be domesticated. A tame animal, with rare exceptions, did not breed in captivity, but domesticated animals by definition have to breed in captivity. Thus, horses are readily domesticated from wild horses, but zebras, despite many efforts to make them domesticated, do not breed naturally in captivity. Bears can be tamed as circus performers, but they too will not breed naturally in captivity.

What puzzled Darwin were the strange associations that were expressed by new variations under domestication. Why were blue-eyed cats invariably deaf? Why did hairless dogs have defective teeth? Why were pigeons with short beaks correlated to small feet?[13] To help him sort out the two categories of variations and their role in both domestication and evolution by natural selection, Darwin chose the rock pigeon, *Columba livia*, as easiest to study. There were many varieties of this species. They were popular among hobbyists in many countries, and they had been domesticated and bred in antiquity. There were at least 20 different varieties that Darwin studied. He believed that if they had been found in the wild, they would have been regarded as different species. Instead, Darwin was convinced that they were all descended from the wild rock pigeon. He crossed a white fantail to a black barb. The offspring were "mottled" in appearance, none looking like either parent. He crossed some of the hybrids to each other and obtained many mottled birds, but one was blue with a white rump and showed double wing bars and barred tail feathers with white edges on them. This was typically that combination of features associated with the wild rock pigeon. To Darwin this was an instance of "reversion to ancestral characters."[14] The belief in Darwin's day was that reversions to an ancestral type (also known as an "atavism") were relatively uncommon but somehow represented a loss of the domesticated features that may have taken centuries to establish. To find such reversions in two generations was striking, and it suggested some sort of causal role for hybridization in bringing about the plasticity of heredity.

Darwin found few breeders of pigeons who believed that the breeds of pigeons were descended from the wild rock pigeon. This belief was also true of breeders of almost all domesticated animals and plants—the prevailing view among them was that many of the ancient breeds were species that had existed in the wild but were now extinct. None made these assertions of a polyphyletic origin on ideological grounds based on religion. They did so because they could not conceive that human selection was capable of bringing about such changes over hundreds or thousands of generations of what Darwin described as "unconscious selection."[15] Using the breeder's terminology, Darwin described the maintenance of breeds as a constant effort of "pulling up the rogues." The rogues were the undesired variations from the desired type. If the color was not right, the size not right, and the shape of the beak not right, those birds would not be bred.[16]

Darwin Proposes a Theory of Pangenesis

During the 1860s, Darwin began work on a book that eventually was published in 1868 as two volumes with the title, *The Variation of Animals and Plants under Domestication.* In his letter to Thomas Henry Huxley (1825–1895) on May 27, 1865, Darwin sent the summary of his "provisional theory of pangenesis" and asked Huxley for his comments,[17] saying that he had thought about "reproduction for many years." Huxley replied, encouraging Darwin to publish, but mentioned that the idea was not new. He particularly urged Darwin to read the 1749 *Natural History* of Georges-Louis Leclerc, Comte de Buffon (1707–1788).[18] Buffon used the term "organic molecules" to refer to minute objects that were formed by eating and that were assembled by an "internal mold" into a form characteristic of the species. His concept is difficult to evaluate in today's terminology, with some thinking of his internal mold as a blueprint or genome for the organism and others thinking of his organic molecules as Darwin-like "gemmules" (the term Darwin used for these inferred transmissible intracellular units) or even genes. Yet others see the organic molecules as building blocks like amino acids or nucleotides. Darwin cited Buffon's work but tried to draw distinctions with his own gemmules.

Darwin argued that the cause of variations was unknown. He cited Prosper Lucas (1805–1885) as claiming variation to be an "aboriginal law" similar to growth or heredity that is universal among living things.[19] Darwin claimed that others speak "of inheritance and variability as equal and antagonistic principles," and he attributed their views to an extension of Lucas's view. Some argued that variability arises from crossing distinct primordial forms. Still others argued that an excess of food, climate change, or exercise is involved in the production of variations. Darwin accepted all types of changes of environment as playing a role in bringing about variation, but he cautioned that the changes are more a reflection of the plasticity of the species or variety and not the specificity of the environment. In that regard, Darwin clearly did not accept a Lamarckian interpretation of inheritance. He cited many examples of bud sports, monstrosities, and sudden varieties appearing that have no correlation to the environment in which they arose.[20] Darwin also did not know whether the alleged greater variation of domesticated forms was true, because he recognized that the "unpracticed eye" sees less variation than the practiced one. He mentions Carl Linnaeus's (1707–1778) great surprise that Lapps could assign a name to each member of a herd of reindeer although to him they just looked like a heap of ants on a hill.[21]

Whether the variations were scant or numerous, Darwin believed that he knew how they arose. He claimed he had a theory to account for the variations: "I venture to advance the hypothesis of Pangenesis which implies that the whole organism, in the sense of every separate atom or unit, reproduces itself. Hence ovules and pollen grains,—the fertilized seed or egg, as well as buds,—include or consist of a multitude of germs thrown off from separate atoms of the organism."[22]

There were many current theories of tiny inferred agents in the world of nature that Darwin could draw on. Darwin was aware of the cell theory proposed in 1838–1839 by Matthais Schleiden (1804–1881) and Theodor Schwann (1810–1882) and its reformulation as the cell doctrine by Rudolf Virchow (1891–1902), who argued that all cells arose from preexisting cells. He was also aware of John Dalton's (1766–1844) atomic theory, which saw molecules as composed of individual atoms that served as elements capable of having affinities for specific other atoms. Darwin referred to these cells or smaller components (including his gemmules) as "minute elements" or centers. He also attributed to them properties such as "latency," in which the trait is not expressed but may be carried along for many generations and then reemerge when a "reversion" to an ancestral type takes place.[23] Darwin also referred to his contemporary, Herbert Spencer (1820–1903), who introduced the phrase "physiological units" for minute substances bigger than known molecules, but smaller than cells, that must play a role in heredity. Darwin attempted to distinguish his gemmules from all these other inferred tiny agents that were associated with the origin of variations.[24]

Critics of Darwin's theory of pangenesis have pointed out that it was largely a rehash of ideas dating back to Hippocrates and that it was not the circulation or even a possible cellular origin that was at issue but the nature of the variations themselves and why they either fluctuated or appeared dramatically, neither of which the pangenesis model addressed. Heredity throughout the 19th century remained a mystery both for its transmission and for the process of mutation or variation. The transmission problem had been solved by Gregor Mendel (1822–1884), but his work was largely ignored as irrelevant to evolutionary variations. For most biologists in the 19th century, the origins of mutations lacked a good model once Lamarck's theory of adaptive response to the environment was repudiated in the 1880s.

But in two respects the 19th century did clarify the mutation process. An experimental test by Francis Galton (coincidentally Darwin's half-cousin) weakened Darwin's pangenesis hypothesis. Galton used blood transfusions in rabbits and showed that silver gray rabbits transfused with blood from white, yellow, black, or gray rabbits did not assimilate any of the alleged gemmules of the donor blood. This was true whether the blood was injected or directly transfused by carotid insertion of a cannula or shunt between the two rabbits. Experiments were also done in the reciprocal direction. The experiments began in 1869 and were presented in the *Proceedings of the Royal Society,* March 30, 1871.[25] Galton used many rabbits for this study and bred the recipients. "From this large stock I have bred eighty-eight rabbits in thirteen litters and in no single case has there been any evidence of alteration of the breed."[26] Galton did not mince words in the implications of his experiments: "[T]he conclusion from this large series of experiments is not to be avoided, that the doctrine of Pangenesis pure and simple, as I have interpreted it, is incorrect."[27]

Although Darwin was aware of the experiments and welcomed Galton's experimental test of the pangenesis theory, he was horrified by the results and protested in print that he never claimed that the gemmules circulated through the blood, but that they could pass from cell to cell by anastomoses between them. It was a lame interpretation that convinced no one, and pangenesis languished after Galton's experiments.

A more powerful argument against pangenesis came from the work of August Weismann, whose articles and books on heredity dominated the last decades of the 19th century.[28] Weismann argued that acquired characteristics would have to pass from somatic to germinal tissue. This was circumvented in Darwin's model of pangenesis by the gemmules. But Weismann claimed that all tests of somatic transmission to the germinal tissue showed no such transmission existed. In humans he cited practices like circumcision, foot binding, head compression, lip piercing, neck stretching, and other practices involving mutilation, stretching, or compression. None of these cultural activities going back in some populations for millennia had ever led to germinal transmission of the trait. Weismann's own experiments amputating the tails of rats for seven generations showed no change in the size of tails among the newborn rats from the amputated parents. Weismann concluded that the alterations by the environment experienced by somatic tissues do not enter the germplasm. He called this "the theory of the germplasm," and it became the prevalent model of heredity from then on. It also led to a search for a mechanism by which variations could arise if the germplasm was shielded from somatic influence.

de Vries Introduces a Model of Intracellular Pangenesis

One such effort was proposed by Hugo de Vries, a Dutch plant physiologist who turned his attention to heredity in the 1880s. In 1889, he published a book that revived pangenesis in a new form. He called his theory and his book *Intracellular Pangenesis.*[29] de Vries shortened the term pangenesis and renamed Darwin's gemmules as "pangenes" (then spelled *pangens*). He rejected the term gemmules for his units because they were too closely associated with Darwin's theory of how they migrated from somatic cells to the germinal material. In de Vries's conception, the pangenes did not move, they remained in the nucleus of the cell.

In the 20 or so years from Darwin's theory of pangenesis in 1868 to the publication of de Vries's model of intracellular pangenesis in 1889, the field of cytology had emerged. Stain technology was introduced in the late 1850s, and chromosomes were observed in the 1870s. In 1868, Ernst Haeckel (1834–1919) identified the nucleus as the site of inheritance because the egg was far more massive than the sperm, the two contained approximately equal-sized nuclei, and they contributed equally to heredity. de Vries accepted that view of nuclear dominance in the cell for the transmission of heredity. He also accepted a growing view among cytologists in the last quarter of the 19th century that the chromosomes served as the site of his pangenes.

de Vries acknowledged priority to Spencer for the idea of physiological units that were larger than known chemical molecules and smaller than cells: "...[T]he material bearers of hereditary characters cannot be identical with the molecules of chemistry. They must be conceived of as units, built up from the latter, much larger than they, and yet invisibly small."[30] Spencer's physiological units, Darwin's gemmules, and many similar units proposed by Carl Nägeli (1817–1891) and August Weismann lacked a mechanism keeping them together and suffered from the Lamarckism that would be associated with them if they freely moved throughout the organism from their cellular location. Weismann partially solved the problem by introducing a concept to interpret the significance of meiosis, partially described by Oscar Hertwig (1849–1922) and Edouard Van Beneden (1846–1910). Weismann proposed a reduction division in germ cell formation to maintain the constancy of number of units associated with a particular species. de Vries cites support for Weismann's theory from contemporaries who compared nuclear size in the reduced (or "haploid," as it was later called) condition and in the nonreduced (or "diploid," as it was later called) condition. Plants having alternations of generations (such as mosses) showed a larger nucleus in the nonreduced or sporophyte stage than in the reduced gametophyte stage. Weismann also proposed that these units could be located in the chromatin, "arranged in rows of the chromatin-thread of the nucleus."[31]

For de Vries, the renaming of gemmules as pangenes was more than a variation on a theme of Darwin's. He took Haeckel's nuclear location for the pangenes and Weismann's chromatin-thread as the site on which they were aligned. He also insisted that they could not leave the cell. They remained in the nucleus until called forth in a particular tissue, and there a selected number of pangenes became active. He believed that they left the nucleus, perhaps when the nuclear envelope dissolved during cell division, but almost never the cell itself. For most of the life cycle of an organism, the pangenes were silent (or latent) in their cells. But what made de Vries's proposal more effective than his predecessors was his recognition of two ways that pangenes could bring about variation. One was to "change their nature" to a plus or minus variation of the character itself, such as intensity of color. The second was if "new kinds of pangens may develop from those already existing" leading to "new characters."[32]

On May 16, 1903, de Vries gave a paper on "Fertilization and Hybridization" to the Dutch Society of Science in Haarlem. In it he proposed a concept similar to crossing over that Thomas Hunt Morgan would later propose for fruit flies harboring several sex-limited traits. In de Vries's mind, "if every unit, that is every inner character or every material bearer of an external peculiarity, forms an entity in the pronucleus, and if the two like units lie opposite each other at any given moment, we may assume a simple exchange of them. Not of all (for that would only make the paternal pronucleus into a maternal one) but of a larger, or even only a small, part. How much and which may be simply left to chance. In this way, all kinds of new combinations of paternal and maternal units may occur in the new two pronuclei, and when these separate at the formation of the sexual cells, each of them will harbor in part paternal and in part maternal units."[33]

To de Vries, then, we can assign three sources of variations from pangenes. The first is the Darwinian plus or minus departures from the standard trait (what Muller in the 1930s would call "hypermorphs" and "hypomorphs"). The second was the appearance of new traits from a gene that did not express such a trait earlier (physiologically, what Muller called "neomorphs"). The third was through a recombination of already existing traits, not through Mendelian production of fixed ratios, but through a process analogous to crossing over in which paired chromosomes (in de Vries's conception, the pronuclei) would exchange components that were aligned in their chromatin-threads. de Vries also recognized that such a mechanism of recombination of paternal and maternal pangenes would greatly increase the variation among offspring. It was not quite the way geneticists would see it a generation later, but it was a substantial advance in the problem of mutation compared to the theories in Darwin's generation.

Throughout the process of the evolution of a concept, there is a letting go and an adding on of ideas and interpretations, as this analysis of hereditary units in the 19th century illustrates. Some of these are associated with the introduction of new facts from new fields. Cytology, for example, was virtually unknown to Darwin and his contemporaries, but it was a major contributor to an understanding of how heredity works in the last quarter of the 19th century.

References and Notes

1. de Vries H. 1901 and 1903. *The mutation theory* (translated into English 1910 by Farmer JC, Darbishire A). Open Court, New York.
2. W.D. Fox to C.R. Darwin, November 1938. *Darwin Correspondence Project,* Letter 418. Cambridge University has published 17 volumes since 1985 covering 6000 letters, with another 9000 eventually to be published. F. Burkhardt is editor-in-chief of the project. The correspondence is available online at http://www.darwinproject.ac.uk. Cited as DCPCU in this list.
3. C.R. Darwin to William Herbert, 26 June 1839. DCPCU, Letter 523.
4. R.S. Ford to C.R. Darwin, 6 May 1839. DCPCU, Letter 509.
5. The 1840 and 1842 sketches are available online, as are his published articles and books. These are on the http://darwin-online.org.uk/ website established in association with Cambridge University Press, which has made library copies available.
6. Darwin F, ed. 1909. *The foundation of the origin of species: Two essays written in 1842 and 1844 by Charles Darwin,* p. 1. Cambridge University Press, Cambridge, UK. (Cited as FOS 1842 or FOS 1844 herein.)
7. Darwin C, FOS 1842, p. 1.
8. Darwin C, FOS 1842, p. 1.
9. Darwin C, FOS 1844, p. 1.
10. Darwin C, FOS 1844, p. 2.
11. Darwin C, FOS 1844, p. 4.
12. Darwin C. 1859. *On the origin of species by means of natural selection,* p. 7. John Murray, London.
13. Ibid., p. 7.
14. Ibid., p. 25.
15. Ibid., p. 35.
16. Ibid., p. 30.
17. C.R. Darwin to T.H. Huxley, 27 May 1865. DCPCU, Letter 4837.
18. Count Buffon (Leclerc GL). 1749. *Natural history* (translated into English by William Smellie and published 1781). Available online at http://faculty.njcu.edu/fmoran/buffonhome.htm/.
19. Darwin C. 1868. *The variation of animals and plants under domestication,* p. 250. John Murray, London. The reference to Prosper Lucas is to his book

Traité philosophique et physiologique de l'héredité naturelle (two volumes, published in 1847 and 1850, respectively).

20. Darwin 1868, op. cit., p. 250.
21. Darwin 1868, op. cit., p. 251.
22. Darwin 1868, op. cit., pp. 357–358.
23. Darwin 1868, op. cit., p. 373.
24. Darwin 1868, op. cit., p. 375.
25. Galton F. 1871. Experiments in pangenesis by breeding from rabbits of a pure variety, into whose circulation blood taken from other varieties had previously been largely transfused. *Proc R Soc Lond B Biol Sci* **19:** 393–410.
26. Ibid., p. 403.
27. Ibid., p. 404.
28. August Weismann's books on heredity (1889) were translated into English in two volumes, *Essays upon heredity and kindred biological problems* (ed. Poulton EB). Clarendon Press, Oxford.
29. de Vries Hugo. 1910. *Intracellular pangenesis* (translated from the German 1889 edition by Stuart Gager C). Open Court Publishers, Chicago.
30. Ibid., p. 49.
31. Ibid., p. 55.
32. Ibid., p. 74.
33. de Vries H. 1903. Fertilization and hybridization. 151st meeting of the Dutch Society of Science in Haarlem (16 May 1903). Op. cit. (note 29), p. 243.

3

Cytological and Mendelian Aspects of Mutation

As THE 19TH CENTURY ENDED AND THE 20TH DAWNED, two events reshaped thinking about variation. The first, in 1900, was the rediscovery of Mendel's two laws of transmission of hereditary traits.[1] That was a continental discovery first announced by Hugo de Vries and independently published by Carl Correns and by Erich von Tschermak-Seysenegg. Although there are ongoing debates on how much each of these three knew of Mendel's work before they wrote up their articles for publication, we can take their own claims at face value. Each thought he had discovered something new until a literature check revealed that what they had actually done was a confirmation of the work published by Gregor Mendel in 1865. de Vries's accounts were most significant because he had extended Mendel's findings to numerous plant species, including commercially valuable crops like maize. This led to a frenzy of studies in agricultural field stations in the United States, mostly for crops like tobacco, beans, tomatoes, squash, maize, and wheat. It also initiated in Europe and North America attempts to replicate Mendelism in animals. Mice, rats, rabbits, and poultry were the first to be explored because of their smaller size and relatively short breeding time. From these new studies, aberrant ratios were found, and this enriched the understanding of variation.

The second major event was stimulated by cytological studies in Continental Europe, especially the elaboration of the processes of cell division—mitosis and meiosis—in the last quarter of the 19th century. The tie of cytology to heredity was most dramatically made in the United States through the work of Edmund Beecher Wilson (1856–1939) and his student Walter Sutton (1877–1916). Wilson associated a specific trait—sex determination—to a specific chromosomal basis and introduced the concept of sex

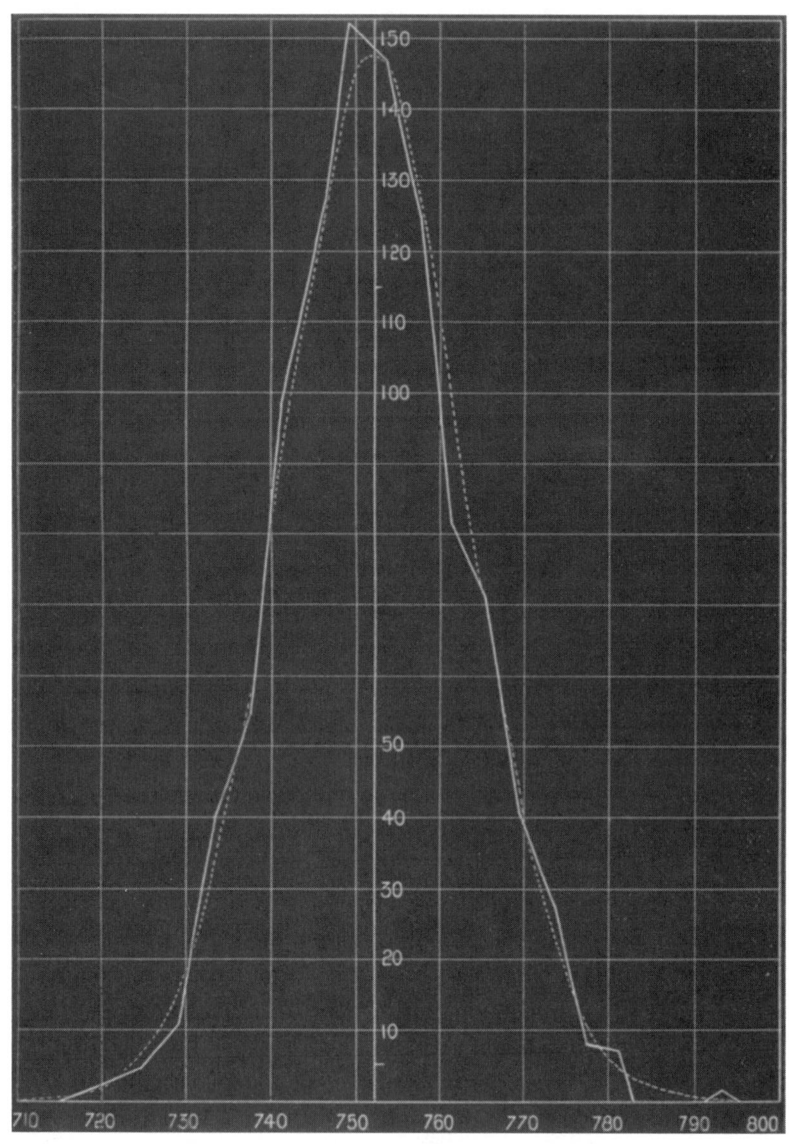

Biometricians use Gaussian curves and their departures for evolutionary analysis. W.F.R. Weldon was a leader of the biometric school that sought Darwinian fluctuations by looking at curves for different populations of animals or plants. In this 1893 study, Weldon studies carapace length in crabs residing in Italy. Bateson at this time was launching his own model of discontinuous variation as the basis for seeking evolutionary change. The rediscovery of Mendelism in 1900 would eventually shift genetics away from the biometric approach. (Reprinted from Weldon WFR. 1893. On certain correlated variations in Carcinus maenus. Proc R Soc Lond **54:** *318–329.)*

chromosomes. Sutton tied the process of meiosis to the distribution of traits in Mendel's laws and showed a perfect correlation, which he called "the chromosome theory of heredity."[2]

The study of variation would take on separate histories in Continental Europe, in Great Britain, and in North America. The emphasis in Continental Europe would be on breeding analysis, with a heavy emphasis on the limits of Mendelian analysis for traits that later were shown to involve cytoplasmic factors. In Great Britain, there would be two decades of bickering by those who felt that Mendelism and cytology were irrelevant to the Darwinian variation associated with natural selection. In North America, a combination of practical uses of Mendelism and a stress on the chromosome theory of heredity would lead to an explosion of interest among universities and field stations in the new field opened up by the rediscovery of Mendelism. German scientists had dominated the field of cytology in the last three decades of the 19th century, but cytology was largely associated with faculty in the medical schools. Great Britain took minor interest in using cytological approaches and stressed evolution, Darwin's influence being central to most academic biology. In North America, the new research PhD degree, introduced at Johns Hopkins University in the 1870s, tended to be interdisciplinary and democratic, with several professors in a single department sometimes supplementing each other's work. This led to a stress on basic research in universities and colleges and applied research, chiefly at the field stations first established in the Abraham Lincoln administration. Cytology flourished at the universities, but not in the medical schools, which tended to ignore advances in genetics as irrelevant to a medical practice. American cytology, unlike German cytology, was primarily a field for the PhD and not the MD.

Mendel Develops Two Laws of Transmission

Mendel's work was extensive. He began his breeding experiments in 1856 and published them nine years later at his local scientific society in Brno (then called Brünn). Mendel was a farmer's son and had chosen the Church as a vocation to attain an education and become a teacher of natural science. His experience on his father's farm left him with a lifelong interest in the varieties of flowering plants he encountered. His education, partly at the University of Vienna, was more concentrated in the physical sciences because that was what he would be teaching. The exposure to combinatorial mathematics, statistics, and the physicist's approach to experimentation served him well when he returned to his monastery and launched his career as a teacher and later as a researcher in his own monastery garden.[3]

While he was in Vienna, Mendel read the work of hybridists, particularly favoring the work of Carl Friedrich Gärtner (1772–1850).[4] Hybrids were of dual interest to Mendel. They represented a source of new varieties of ornamental flowers, an interest that went back to his childhood on the farm, and they were a key to studying the species problem. He and his teachers were familiar with the work of earlier hybridists in Europe and Great Britain, especially Gärtner, Josef Gottlieb Kölreuter (1733–1806), William Herbert (1778–1847), Henri Lecoq (1802–1871), and Max Ernst Wichura (1817–1866).[5] From their work Mendel learned that some crosses between varieties yielded sterile offspring. Other crosses yielded robust offspring. Hybrids tended to be unstable if fertile and to yield a variety of offspring when bred to one another. In his published account of his experiments, Mendel stated that his objective in doing them was to look for specific kinds of varieties and their numerical proportions "to ascertain their statistical relations." He cited the underlying rationale for his studies was "the history of the evolution of organic forms."[6]

The phrase Mendel uses for the variations he studied is "constant differentiating characteristics."[7] Mendel claimed that he bred 34 varieties of peas for consistency of their expression and selected 22 of them for further breeding. He then narrowed this batch to seven traits in each of which one form was complementary to a corresponding form, such as color of the cotyledons (yellow vs. green) or the shape of the pod (inflated rather than collapsed around the peas within them). He examined 10,000 plants derived from these crosses and found very few cases of accidental pollination (probably by the pea weevil, *Bruchus pisi,* or *Bruchus pisorum*). He had chosen peas because their flowers were largely self-fertilizing and well protected from accidental cross-pollination. The cross-pollination Mendel did between varieties was by hand in which he would cut off the stamens and then dust the pistils with pollen from a desired type.[8]

Mendel introduced two terms that would change the way varieties were described in future research in this field. He noted that in the hybrid of each of the seven pairs of traits he studied, the hybrid expressed only one of the parental contrasting traits. He called this expressed trait "dominant," and he designated the latent or unexpressed form as "recessive" because it reappears later when the hybrids are allowed to self-fertilize. He also pointed out that it made no difference whether the trait that was dominant came from the pollen or the ovule. Reciprocal crosses gave identical ratios and results.[9]

Mendel's paper is carefully constructed to be descriptive of the traits he used. He avoids speculating about the underlying basis for their expression. He does not use terms like "physiological units" or draw analogies to atoms. The closest he comes is the use of the vague term "elements" when he describes the fertilization process as a "material composition and arrangement of the elements which meet in the cell in a vivifying union."[10]

Mendel recognized that large numbers of such "elements" could give complex ratios and numerous new varieties. His mathematical approach used a combinatorial approach to generate predicted ratios from two or three factor combinations, and his statistical approach used the approximately 3:1 ratio (for one pair of traits, three expressing a dominant trait and one the extracted recessive) to become a law or rule of a fixed theoretical ratio. Similarly, when the dihybrid was selfed to generate a 9:3:3:1 ratio of the double dominant, the two single dominants, and the double recessive, Mendel used large enough samples from his many crosses to show a consistency with the theoretical ratio. Unfortunately, those in attendance at the February 8 and March 8, 1865, meetings in Brünn as well as those who read his article in the *Proceedings of the Natural History Society of Brünn* were not excited by this new approach. Many rejected it as a type of "numerology" that did not connect to their more pragmatic approach that avoided laws in biology.[11] As far as is known, neither Darwin nor Galton read Mendel's paper, and very likely Darwin would have rejected it as not significant for his theory of natural selection. Mendel's traits were more suitable for following the fate of sports and other pathologies and not the small fluctuating differences Darwin favored for evolutionary change.

Darwin's Supporters Favor a Biometric Approach to Evolution and Variation

A field-based quantitative approach to heredity rather than experimentation was certainly the way Darwin's intellectual heirs saw it in the last quarter of the 19th century. They developed what became known as the "biometric school," and their three chief advocates were Galton, Karl Pearson, and W.F.R. Weldon. Their approach was largely statistical rather than combinatorial. They measured change through fieldwork, carefully measuring lengths of appendages and other traits in populations, and they produced curves, many of them Gaussian normal curves and some showing unusual peaks that reflected selective events in which two populations contended in a common territory. Galton was their theoretician, and although he was open to all types of models of heredity, he favored the biometric school's efforts.[12] Although fluctuations as Darwin proposed had no known cause or mechanical model that could bring such subtle changes into being, Galton in his book *Natural Inheritance* (1889) had much more insight into mechanical modeling for the origin of variations and sports.[13] He saw the Darwinian variations as outcomes of a polygon-like crystal with several facets. If the environment disturbed the polygon, it would rock or vibrate and thus give plus or minus variations. More dramatic departures from normalcy occurred if his crystal-like object with facets was resting in a stable position but could occasionally be tipped over by some environmental trauma into an unstable position or facet, resulting in a sport. He called this a "polygon of instabilities."[14] Like Darwin, Galton did not consider the sports as having evolutionary significance.

Bateson Advocates Discontinuous Variations for Evolutionary Change

William Bateson's views of variation were quite the opposite. Bateson began his career in the life sciences studying embryology at Cambridge. He spent a useful and productive time in Chesapeake Bay and Johns Hopkins University working with William Keith Brooks (1848–1908), who was writing a book on heredity. Brooks believed "the problem is two-sided; what is now heredity was at one time variation, and each new variation may soon be hereditary. Heredity and variation are opposite aspects of the same thing. . . ."[15]

Brooks convinced Bateson that a study of heredity would prove fruitful if it were shifted from speculation to experimental science. Bateson considered that sound advice. Bateson first completed an evolutionary project on the origin of vertebrates by studying the embryology of a hemichordate, the acorn worm *Balanoglossus*, and relating its larval notochord to vertebrate characteristics. He returned to Cambridge and then studied variation by poring through library references to abnormalities of all kinds. He added two new ideas about hereditary variations. In addition to Darwinian fluctuations, which he considered unsatisfactory for explaining the origin of higher categories of classification (phyla, especially), he classified some sports as "meristic variations" if they involved increases or decreases in parts such as digits of hands or feet, teeth, or vertebrae.[16] Much of an organism's early development, he argued, is meristic and consists of duplications of parts like the somites in vertebrate embryos. A second category of sports he reconceptualized with the term "homeosis."[17] He believed that such "homeotic variations" played a significant role in the evolution of higher taxonomic categories. These variations involved the displacement of a developmental unit from one organ system to another. The presence of a tuft of feathers, where a coxcomb would normally be in domesticated chickens, was one such example that could lead to species differences because the homeotic organ could differentiate through selection into a new functional role. In nature, such a modification of a tuft of feathers had become established in South African secretary birds. The third classification of hereditary sports that Bateson offered was one having a pronounced effect on body plans. These he called "symmetric variations" in which bilateralism could be converted into radial symmetry or other geometric variations.[18]

Bateson collected this resulting scholarly survey of the literature on variations in both natural and domesticated plants and animals into a book, *Materials for the Study of Variation* (1894). The book is written in an argumentative way that no doubt alienated the biometricians because it was a throwing down of a gauntlet, charging them with being on the wrong side of the way evolution should be studied. He rejected the term "heredity" because it implies "property"; instead, he preferred to stay with the terms "continuous variation" or "discontinuous variation," which are descriptive rather than interpretive. He chastised field biologists for not paying heed to the variations they find and instead focusing on the typical or expected characteristics for systematic description. Bateson disliked the term "reversion," especially to an ancestral form, because the attributes in the alleged cases are vague or sparse, and he denied that "useless" parts of an organism are more variable than vital components.[19] He claimed that his book was written not to supply answers but to present problems and the materials needed to study them. Bateson also dismissed the environment as playing any significant role in producing the types of variations he described: "Discontinuity . . . has its origin not in the environment nor in any phenomenon of Adaptation, but in the intrinsic nature of organisms themselves, manifested in the original Discontinuity of Variation."[20]

For Bateson, meristic variations arise because "the bodies of living things are mostly made up of repeated parts—or organs or groups of organs—which exhibit the property of 'unity' or, as it is generally called, 'individuality'."[21] Finally, Bateson rejected the idea of "swamping," a concern that shook Darwin when it was first proposed. In swamping, any new variation would be blended out of existence by reproduction, and it would enter a latency or remain outnumbered by the greater number of variations already present in the individual or population. This would not happen if the significant traits for evolution were discontinuous. They would be selected if they had a beneficial effect on the progeny. Bateson cited many examples of discontinuous traits (sports) that did not fade out or get swamped, as they were bred generation after generation to other varieties, especially to those with normal traits associated with a breed or species.[22]

Bateson proved prophetic when he concluded his study of variation, "The only way in which we may hope to get at the truth is by the organization of systematic experiments in breeding, a class of research that calls perhaps for more patience and more resources than any other form of biological inquiry. Sooner or later such investigation will be undertaken and then we shall begin to know."[23] Whatever ill feeling existed between Bateson and the biometricians when he came back from his association with Brooks, things became intensely bitter after the publication of his massive treatise on variations. For the next 30 years, he would encounter ridicule, ostracism, or fierce rejection of his views on heredity. His opponents were as blunt as he in presenting and defending their views and findings. Sadly, this argumentative style worked against both camps. For the biometricians, it led to their being marginalized as the new movement based on the chromosome theory of inheritance penetrated to most of the world. Outsiders, rather than the biometric group, would found the field of population genetics. For Bateson, it meant that his belligerence would outlast the contributions he made to the field he named, genetics, and he lapsed into an underrated figure whose ideas were revived some 50 years later but without attribution to him.[24]

The Rediscovery of Mendelism Launches 20th Century Mutation Studies

What changed the study of variation was the rediscovery of Mendelism. Bateson immediately embarked on the program he recommended and built a productive school of students and research associates who carried out both plant and animal breeding experiments.[25] This led to new phenomena and new terminology in a scant six years. Bateson assigned a new term to replace Mendel's phrase "differentiating characters." He called them "allelomorphs" (later changed to "alleles"). This was an important conceptualization. It suggested that there was a place or location for the two elements or units that allowed their association in pairs, an essential feature if the hybrid with two alleles sorted the two alleles into different germ cells. He took Galton's human pedigree notation of P_1, F_1, and F_2 and used these to represent the parents', children's, and grandchildren's generations for any organism. He introduced the concept that alleles could be "homozygous" as in **AA** or **aa** or "heterozygous" as in **Aa**—those terms and symbols had a powerful effect on rendering traits into mental abstract units that could be manipulated on paper to design experimental crosses and to organize data from crosses. The term "heterozygous" also purged older terms such as "hybrid," "bastard," "mongrel," or "half-breed" of their odious connotations. He also named the new field "genetics," which at once (even if not intentionally) incorporated the pangenes or hereditary units into heredity. Instead of inheriting property, which would have implied in those days a male transmission in the minds of those who received benefits from an estate, it implied receiving the elements of heredity themselves from both parents. A few years later, Wilhelm L. Johannsen (1857–1927) trimmed the term "pangene" and shortened it to "gene," an otherwise undefined unit of transmission. Bateson also recognized from his breeding experiments that the term "dominance" could be modified, some traits showing "complete dominance" and others showing "partial dominance."[26] Bateson erred in calling Mendel's elements "unit-characters or unit factors." This bound too closely an agent of transmission with the expression of a trait, creating confusion for the recessive traits. That confusion was resolved when Johannsen used the term "genotype" for the inferred genetic composition of an organism and the term "phenotype" for the appearance of a trait.

Most of Bateson's new terminology appeared in a set of *Reports to the Evolution Committee of the Royal Society (1902–1906)*, a committee that Galton had set up with both Bateson and Weldon as members. Galton hoped that that they would work together and iron out their differences as committee members. Instead, Weldon boycotted the meetings after some heated exchanges with Bateson. This allowed Bateson and his colleagues to publish their new experimental studies using breeding analysis to identify new modified Mendelian ratios, interactions between different pairs of alleles, and later the discovery of linked associations of these traits. Breeding analysis was widely adopted in universities and research institutions around the world, and commercial applications quickly followed. Genetics became an attractive science to the agricultural industry and one of the rising stars in theoretical biology because of its new ways of describing life.

The Chromosome Theory of Heredity Appears in the United States through the Work of Wilson and His Students

The chromosome theory of heredity came from the confluence of cytology and breeding analysis in the United States through the work of Edmund

Beecher Wilson and his students. In Germany it came through the confluence of cytology and experimental embryology (*Entwicklungsmechanik*), particularly through the work of Theodor Boveri (1862–1915).[27] Wilson's team followed the debates on a finding by Hermann Henking (1858–1942) in Germany in 1890 that *Pyrrhocoris aptera*, a fire wasp, had in its cell nucleus an unusual object that was either an unusual chromosome or an unusual nucleolus. Because Henking did not know what it was, he gave it the symbol for mathematical unknowns and called it an "X element."[28] Later Clarence McClung found what he called an "accessory chromosome," and it was correlated to males and not females in a grasshopper he studied.[29] Wilson began a study of Hemipteran bugs, and Nettie Stevens began a study of Diptera and other insects to study these differences in chromosomes between males and females.[30] Independently, Stevens and Wilson found a situation in which a pair of chromosomes differed in the two sexes. What Stevens called "heterochromosomes," Wilson initially called "idiochromosomes" and later "sex chromosomes," which he symbolized by X and Y. This made females XX and males XY in many of the insect species that they studied. The story rapidly became complex as some bugs had XO (no chromosome accompanying the X chromosome at meiosis) as male and XX as female. Still others had multiple Y chromosomes. What they shared in common was that there were "sex chromosomes" and a shared set of "autosomes" or chromosomes found in pairs in both males and females.[31] For Wilson, sex was a specific trait that could be assigned a chromosomal basis.

Wilson's student Walter Sutton took an interest in these chromosome studies, and he discussed with Wilson the implications of the newly rediscovered Mendelian laws. Wilson urged Sutton to publish his idea that there was a correlation between the reduction division and the distribution of heterozygous alleles to separate gametes. Fertilization brought these together and resulted in the laws Mendel had first demonstrated. This was also true for the pairs of chromosomes associated with different pairs of alleles. Their independent assortment due to random alignment of the homologous chromosomes served as a mechanical model for the distribution of gametes that, when merged in fertilization, led to the 9:3:3:1 ratio of Mendel. Sutton recognized that even if an organism had about 30 chromosomes, there would be millions of combinations of alleles if each chromosome harbored a heterozygous pair. Sutton called this the chromosome theory of heredity, and he published his paper in 1903.[32]

About the same time, in Germany, Theodor Boveri used mechanical agitation to form multiple spindles in fertilized eggs of sea urchins, which produced blastomere nuclei with different numbers of chromosomes. These isolated blastomeres formed abnormal larvae in which the abnormality of organs and size of larvae in each case were associated with a particular abnormal number or kind of chromosome they contained in excess or deficit. This demonstrated that chromosomes differed qualitatively in their contributions to the development of the organism. Wilson dedicated his 1896 book on the cell to Boveri.[33] Although Boveri's work had a profound effect on Wilson and his students, they differed in their approaches. Boveri did not associate specific Mendelian mutations or traits to specific chromosomes. The sex chromosome model did. For those reading Sutton's paper, a possible assignment of genes to chromosomes was not far off, and it would use Mendelian genetics to do this. It was more difficult using Boveri's experimental embryology approach to bring about that union of specific inherited traits and specific chromosomes. Nevertheless, some historians refer to the chromosome theory as the Sutton–Boveri theory. This unfortunately leaves out Wilson's significant contributions, and therefore I would recommend calling it the Sutton–Wilson–Boveri chromosome theory of heredity.[34]

de Vries Loses Interest in Pursuing Mendelism

While working in 1899 on his plant breeding experiments in Hilversum (The Netherlands), de Vries noted something unusual in a field of evening primroses. The normal traits associated with *Oenothera lamarckiana* seemed dramatically altered in several ways in a new variety. It did not breed with its parental type, but it did self-fertilize, and in this form it bred true, producing only plants of the new variety. To de Vries this was the production of a new species and he gave it a name, *Oenothera gigas*. Over the next few years, de Vries found additional new species as well as varieties that did breed with either other new species or with the parental *O. lamarckiana*. de Vries believed that he had encountered a mechanism of evolution through sudden jumps (called "saltations"), and thus the evolution of new species was not gradual as Darwin conceived it in his theory of natural selection, but sudden, with discontinuity rather than continuity marking the generation time between the old and the new species. He called his findings "the mutation theory" of evolution and gathered his data and writings into a two-volume work that he called *The Mutation Theory*.[35]

The biometrical school in Great Britain was not impressed by de Vries's contribution. Bateson was initially elated, but he realized that there was something odd about de Vries's plants because no such sudden appearances had occurred among the tens of thousands of specimens he had counted and observed over five years of breeding. Bateson felt that the emphasis should be on the discontinuity of individual traits and their role in evolution and not a discontinuity in which the heredity en masse undergoes a change. But exactly that idea appealed to Thomas Hunt Morgan. He visited de Vries in Holland about 1900 and came away with enthusiasm for what he believed to be an experimental approach to the study of evolution.

What Bateson, de Vries, and Morgan shared was an interest in discontinuity in the evolution of organisms. Each chose a different way to pursue that interest. Bateson stayed with Mendelian varieties and never found a way to plug those findings into evolution, leading to his pessimism and doubts in his later years. Perhaps it was his dislike of the biometricians that blinded him to the new findings in population genetics that emerged while he was still alive. For de Vries, it was the conviction that his mutation theory of evolution was going to be the new Darwinian revolution and that he had found the mechanism by which new species arise. He introduced terms like "mutating periods" to account for the richness of new species and varieties in his primroses but their relative absence in other species. He too became bitter as his work slowly eroded and other geneticists either lost interest in his model or began to identify the mechanisms that created his alleged new species, demoted over the next two decades to varieties of *O. lamarckiana* having an assortment of different chromosomal aberrations.[36]

While de Vries' mutation theory faded out of memory among a new generation of classical geneticists, the term "mutation" that he introduced shifted its meaning to events in the individual Mendelian units or genes. The winner in this triumvirate was Morgan. He adopted the chromosome theory of heredity and applied it to evolution after shedding his initial skepticism about it, and he abandoned de Vries' mutation theory after initially embracing it and putting it to a test. Both Bateson and de Vries appear to have had personalities such that they were wedded to their original ideas. Morgan, in contrast, made his reputation as a skeptic willing to speculate and to abandon ideas using experimental tests as the only way to convince himself or others that his ideas were worthwhile.

Morgan Founds the Fruit Fly School of Genetics

Morgan, like Bateson, found the experimental approach and the enthusiasm for new knowledge at Johns Hopkins University to be invigorating. He liked Foster's (Michael Foster [1836–1907]) experiments in physiology, and he liked using a microscope and collecting specimens in Brook's movable summer station set up at different locations in the Chesapeake Bay. Morgan's PhD was on the embryology of sea spiders (pycnogonids), and after receiving his PhD, Morgan sought the stimulus of European scientists, especially those who frequented the Marine Biology Station at Naples. Morgan loved Italy because his father had been a State Department consul in Italy before the Civil War. Morgan developed an outlook that few substantial advances in the life sciences were possible without doing experiments on living things. This was reinforced by the successes of German biologists like Wilhelm Roux (1850–1924), Hans Driesch (1867–1941), and Boveri who used numerous techniques to manipulate cells to study events in cell division and the early stages of embryonic development.[37] Morgan found that he could do the reverse of twinning and fuse two fertilized eggs or early blastulae together and produce one chimeric individual. He began a study of problems of regeneration of lost limbs and other body parts, hoping that these would lead to insights about heredity. He studied parthenogenesis in aphids and phylloxerans and hoped that would lead to ideas about sex determination. On one sabbatical leave, he went to Holland and discussed the origin of species by mutations with de Vries, who had just begun publishing his mutation theory that he hoped would replace evolution by Darwinian fluctuating variations. Morgan avoided the speculative approach to interpreting evolution, heredity, or sex determination, but he was not afraid to generate tentative hypotheses as long as they could lead to experimental tests.

About 1904, at his relatively new position at Columbia University, Morgan began a series of experiments looking for a "mutating period" in mice, poultry, pigeons, hamsters, and other animals. He realized quickly that these organisms were expensive to maintain, difficult to produce in large numbers without financial support, and provided too few progeny compared to the primroses that de Vries was using. In a conversation that Morgan had with William Castle (1867–1962), who was at Harvard, Castle recommended using fruit flies. Castle had just begun publishing on fruit flies in 1905, describing selection experiments for characteristics like wing vein variation and other Darwinian minor differences. The advantages of flies, he argued, were their rapid turnaround time per generation (about 10 days), their large number of offspring per pair of parents (about 300), and the low cost of feeding them (mashed Concord grapes or mashed bananas as a food pressed into the bottom of a small cream jar).[38]

Morgan began his fruit fly studies in 1907. He asked his student Fernandus Payne to collect some for him, and Payne began a project testing degeneracy in fruit fly vision by isolating fruit flies from light (eventually for 69 generations).[39] At the same time, Morgan began his tests of speciation by new mutations, hoping that the large numbers he would produce might give some new species of flies or at least variations that he could study. By 1909, Morgan had little to show for his effort. He found one trait that produced a trident-shaped streaking of the thorax (he called this variation "with") in contrast to the normal unstreaked thorax (he called that trait "without"). He also found a variation that produced a puddling of pigment where the wings joined the thorax, and he called that trait "speck."[40] They were somewhat Mendelian, especially speck. The trident pattern was subject to minor Darwinian variations in expression and could be selected for darker or lighter or sparser presence, and hence it behaved as both a Mendelian trait (recessive) and a variable trait. The speck trait was more stable and acted like a simple recessive. Neither was impressive compared to the dramatic multicharacter mutations that arose as sports in *Oenothera*. Morgan's luck turned in January 1910, when he found a white-eyed male among one bottle of flies. He isolated the male, mated him with several of his sisters, and obtained several white-eyed males and a few white-eyed females. From this new variation, he established a white-eyed stock and bred white-eyed flies and red-eyed flies in reciprocal crosses. He quickly found a departure from Mendelian inheritance. White-eyed males crossed to red-eyed females from other bottles gave only red-eyed F_1 offspring. The $F_1 \times F_1$ cross gave a 3 red to 1 white ratio, but all the white-eyed flies were males, two-thirds of the red-eyed flies were females, and one-third were males.[41]

Morgan had discussed sex chromosomes many times with his Department Chair, Wilson, but he was not convinced that the case for a chromosomal basis for these "sex-limited" traits was proven for his fruit flies. Morgan knew that it was not universal because it had long been known that bees were diploid for females and haploid for male drones. But Morgan soon found two new mutations that behaved the same way as white eyes—miniature wings and rudimentary wings. With three traits all exhibiting sex-limited inheritance, it seemed reasonable to Morgan to assign all three to the X chromosome. He also got a bonus when he crossed these to one another. He obtained evidence of linkage and interpreted that linkage by assuming that paired X chromosomes undergo a crossing over and the resulting X chromosomes were partially paternal and partially maternal in their gene content.[42]

Morgan realized that there were far more new events happening than he had time to pursue, and he began adding students to his laboratory, especially when Payne chose to do his PhD with Wilson instead. He also began to have doubts that what he was observing in flies was a mutating period. The kinds of changes he was obtaining were not species-forming. The mutations he was obtaining were largely Mendelian or at least consistent with changes from a normal allele to a mutant allele.

References and Notes

1. For the original language and English translation of Mendel's paper and 26 documents published between 1898 and 1905 of the rediscovery of Mendel's laws, see Krizenecky J. 1966. *Fundamenta genetica.* Czechoslovak Academy of Sciences, Prague.
2. For an account of this development, see Chapter 7, "Sex chromosomes," in Carlson EA. 2004. *Mendel's legacy: The origin of classical genetics.* Cold Spring Harbor Laboratory Press, Cold Spring Harbor, NY.
3. Iltis H. 1932. *Life of Mendel.* Norton, New York. Orel V. 1996. *Gregor Mendel: The first geneticist.* Oxford University Press, Oxford.
4. Gärtner's work is discussed in Robert HF. 1929. *Plant hybridization before Mendel,* pp. 164–168. Princeton University Press, Princeton, NJ. Gärtner worked with more than 700 species seeking hybrids and did about 10,000 crosses in a 25-year study published in 1839.
5. Olby R. 1966. *Origins of Mendelism.* Constable, London.
6. Mendel G. 1865. Experiments in plant hybridization read 8 February and 8 March 1865, Natural History Society of Brünn. In Krizenecky 1966 (note 1), op. cit., p. 1. Mendel G. 1866. Versuche über Plfanzenhybriden. Verhandlungen des naturforschenden Vereines in Brünn, Bd IV für das Jahr 1865, Abhandlungen, 3–47. Mendel's paper was first translated into English by C.T. Dreury for William Bateson in 1901 and published by the Royal Horticultural Society for its meeting that year.
7. Mendel 1865 (note 6), op. cit., p. 2.
8. Ibid., p. 2.
9. Ibid., p. 4.
10. Ibid., p. 4.
11. Ibid., p. 11.
12. For a detailed account of Bateson's battles with the biometric group, see Cock AG, Forsdyke DR. 2008. *Treasure your exceptions: The life and science of William Bateson.* Springer, New York.
13. Galton F. 1889. *Natural inheritance.* Macmillan, London.
14. Ibid., p. 28.

15. Brooks WK. 1883. *The law of heredity: A study of the cause of variation and the origin of living organisms,* p. 10. John Murphy, Baltimore. Brooks believed that cells regulate their own gemmule production and accumulation. He also had a curious idea that males produce more gemmules than females. What impressed Bateson, however, was Brooks's insistence that experimentation and observation, not theorizing, would solve the problem of heredity.
16. Bateson W. 1894. *Materials for the study of variations: Treated with special regard to discontinuity in the origin of species,* p. 568. Macmillan, London.
17. Ibid., p. 570.
18. Ibid., p. 569.
19. Ibid., p. 78.
20. Ibid., p. 567.
21. Ibid., p. 568.
22. Ibid., p. 573.
23. Ibid., p. 574.
24. Op. cit. (note 12).
25. These were mostly published in a series of volumes called *Reports to the Evolution Committee of the Royal Society 1902–1906.* They are included in a two-volume collection of his published papers: Punnett RC. 1928. *Scientific papers of William Bateson.* Cambridge University Press, Cambridge. (Cited as Reports herein.)
26. Reports 1904, p. 130.
27. Boveri T. 1902. Über mehrpolige Mitosen als Mittel zur Analyse des Zellkerns. *Verh D Phys-Med Ges (Würzburg NF)* **35:** 67–90.
28. Henking H. 1891. Über Spermatogenese und derem Bezeihung zur Entwicklung bei *Pyrrhocoris apterus. Z Wiss Zool* **51:** 685–736.
29. McClung CE. 1899. A peculiar nuclear element in the male reproductive cells of insects. *Zool Bull* **2:** 187–197.
30. Stevens NM. 1905. *Studies in spermatogenesis with special reference to the accessory chromosome,* Publication No. 36. Carnegie Institution, Washington, DC.
31. Carlson EA. 2008. The sex chromosomes. In *Mendel's legacy: The origin of classical genetics,* Chapter 7. Cold Spring Harbor Laboratory Press, Cold Spring Harbor, NY.
32. Sutton WS. 1903. The chromosomes in heredity. *Biol Bull* **4:** 231–251.
33. Wilson EB. 1896. *The cell in development and inheritance.* Macmillan, New York.
34. Op. cit. (note 27). For a biography of Boveri, see Baltzer F. 1967. *Theodor Boveri: The life of a great biologist 1862–1915* (translated from the German by Rudnick D). University of California Press, Berkeley, CA.
35. de Vries H. 1901 and 1903. *The mutation theory* (translated into English 1910 by Farmer JC, Darbishire A). Open Court, New York.

36. Cleland RE. 1972. Oenothera *cytogenetics and evolution.* Academic Press, New York. In this monograph, Ralph Erskine Cleland (1892–1971) gives a comprehensive history of *Oenothera* from its initial promotion by de Vries for his mutation theory to the writing of this encyclopedic work. Many contributed to the field, but Cleland discovered the linked-ring translocations that gave rise to most of the species and varieties that de Vries thought he had found.

37. There are two biographies of Thomas Hunt Morgan: Shine IB, Wrubel S. 1976. *Thomas Hunt Morgan: Pioneer of genetics.* University Press of Kentucky, Lexington, KY. This book covers much of his personality and youth in Kentucky. The other—Allen GE. 1978. *Thomas Hunt Morgan: The man and his science.* Princeton University Press, Princeton, NJ—is a comprehensive biography that goes into detail on his scientific contributions as an embryologist and as a geneticist.

38. I had interviewed Payne at Indiana University, and he provided this background to Morgan's work habits, his personality, and the circumstances that led to the first use of fruit flies at Columbia University. Carlson papers, Cold Spring Harbor Library Archives.

39. Payne F. 1911. *Drosophila ampelophila Loew* bred in the dark for sixty nine generations. *Biol Bull* **21:** 297–301.

40. See Morgan TH, Bridges CB. 1916. Sex-linked inheritance in *Drosophila,* Publication No. 237. Carnegie Institution, Washington, DC.

41. Morgan TH. 1910. Sex limited inheritance in *Drosophila. Science* **32:** 120–122. Also see Green MM. 1996. The genesis of the white-eyed mutant in *Drosophila melanogaster*: A reappraisal. In *Perspectives in genetics: Anecdotal, historical, and critical commentaries 1987–1998* (ed. Crow JF, Dove WF), pp. 504–505. University of Wisconsin Press, Madison, WI.

42. Morgan TH. 1911. Chromosomes and associative inheritance. *Science* **34:** 636–638.

4

The Fly Lab Redefines Mutation

THOMAS HUNT MORGAN'S EARLY CONTRIBUTIONS to mutation were primarily from the work he did from 1908 to 1912. His first fruit fly publications began in 1910. He thought at first that he had found a "mutating period," which undoubtedly must have cheered Hugo de Vries when Morgan's article appeared.[1] But it was soon apparent to Morgan that the novelty of sex-limited inheritance, a new finding, was far more interesting than just a flurry of difficult-to-analyze mutations of no particular evolutionary significance. Getting another two sex-limited traits—rudimentary and miniature wings—was the key to a dramatic shift in his thinking.[2] Ever the skeptic, Morgan had initially resisted Edmund Beecher Wilson's enthusiasm for the idea of the X chromosome as the basis for sex-limited inheritance. The evidence was not the existence of white-eyed flies with a peculiar 3:1 ratio. That was just a correlation. But three new occurrences of sex-limited traits showed the same behavior. When crossed to one another, they produced strange percentages that were like William Bateson's reduplication series in flowering plants, which Bateson could not adequately explain. Bateson proposed a "reduplication model" with his unit characters multiplying within the germ cells, resulting in these atypical ratios that had little predictive value and that demanded a virtually untestable capacity for differential multiplication and for adhesiveness. Bateson used the term "coupling" to describe the occasions when the traits were together in one gamete and "repulsion" when they stayed apart in another gamete. The associations were fixed for any dihybrid he tested, and Bateson's terms implied some sort of magnetic or adhesive mechanism.[3]

Instead, Morgan argued that the dihybrid sex-limited factors he studied were more likely associated with the X chromosome and aligned along its length. Crossing over could bring them together or separate them. In general,

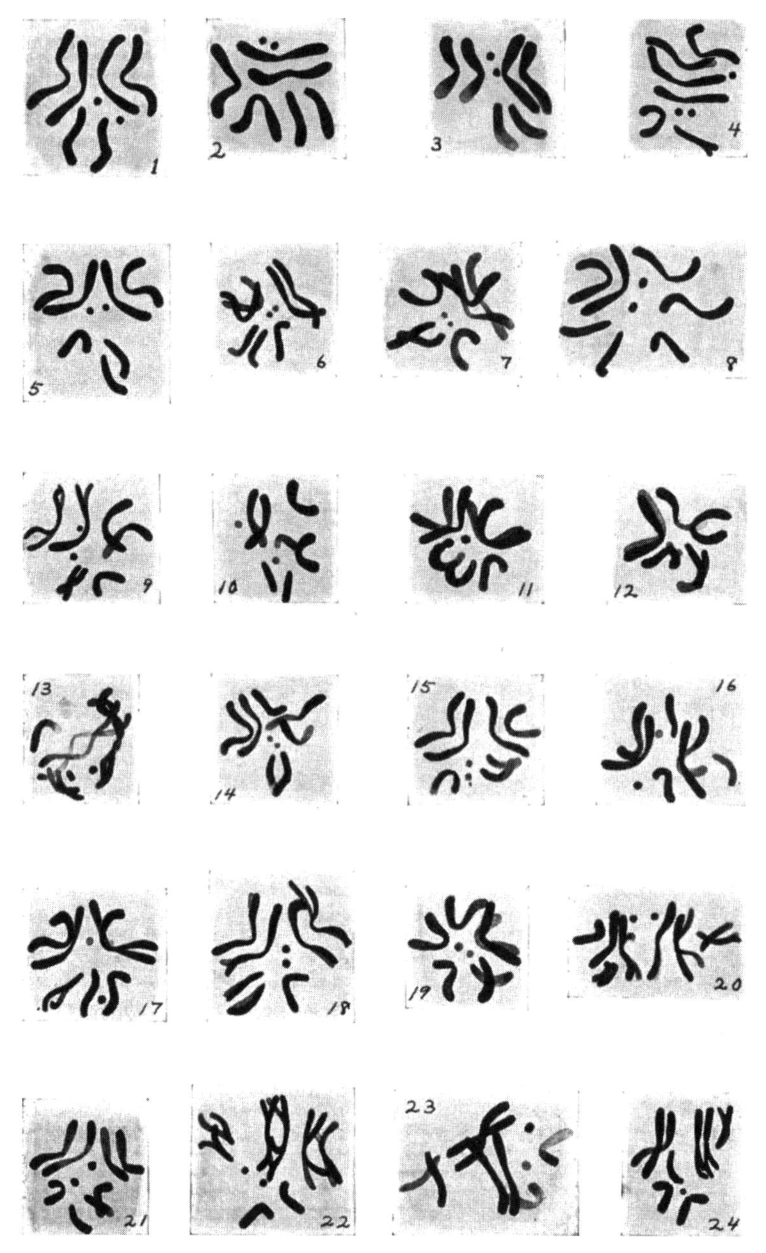

Calvin Bridges identifies extra or missing sex chromosomes in fruit flies. This photographic display appears in Bridges's 1916 publication "Non-disjunction as Proof of the Chromosome Theory of Heredity." The normal chromosome number in somatic tissue is 8. In number 1 the second and third chromosomes are on top and the two X chromosomes are on the bottom. The two dots in the middle are for chromosome 4. In number 5 the second and third chromosomes are on top, the two fourth chromosomes in the middle, and a Y chromosome is in the lower left and two X chromosomes are on the lower right. Bridges related chromosome number to phenotype of eye color using crosses of red- and white-eyed flies with or without non-disjunctional history. (Reprinted from Bridges CB. 1916. Non-disjunction as proof of the chromosome therapy of heredity (concluded). Genetics **1:** *107–163.)*

those factors close to one another would not undergo much crossing over, and those that were far apart would readily undergo crossing over.[4] It was a mechanical model that at least had the support of the "chiasmatypie" of Frans Alfons Janssens (1865–1924) that inspired Morgan's model. Janssens, a Belgian biologist, noted unusual twisting of chromatids in the paired chromosomes of salamanders undergoing meiosis during reduction division.[5] Morgan assumed that the tension of the twists would be relieved by breakage, and a subsequent reunion of a maternal and a paternal broken end would produce a recombinant chromosome. Janssens hypothesized that the chiasmata held the pairs of homologs together (in a stage called the diplotene of reduction division), and breakage allowed them to reconstitute and separate to the daughter cells.

Morgan began adding technicians and students as the work increased. We associate what became known as the "fly lab" at Schermerhorn Hall at Columbia University with the most famous of his students—Alfred Henry Sturtevant (1891–1970), Calvin Blackman Bridges (1889–1938), and Hermann Joseph Muller. A larger number, several getting PhDs with Morgan, failed to make the grade in major discoveries and they have largely been forgotten.[6] Sturtevant was the first to make a major discovery. He used Morgan's data to construct a map of the six known sex-limited traits Morgan mentioned in his class.[7] This included vermilion eye color, eosin eye color, and yellow body color. It was the first map of hereditary factors, and, as Muller described it to me, "it was like a bombshell" in its effect on his thinking. In Sturtevant's map, eosin mapped to the same site as white eyes. Both white and eosin were very close to yellow body color. Vermilion was between white eyes and miniature wing. The farthest mutation Sturtevant mapped was rudimentary wing. By constructing a linear map, Sturtevant represented the chromosome as a line—mathematically, a line is composed of points. Thus, the Mendelian factors or genes became points on a line. In Sturtevant's map, yellow is near the terminus of the map for the X chromosome, and rudimentary is at the opposite end of the set of six traits on the X chromosome. That representation was useful for predicting where new mutations would be located and how much effort was needed to construct genetic stocks that combined two or more of these factors. The final conceptual change by 1913 was that Morgan's descriptive term "sex-limited traits" was being replaced by the more interpretive term "X-linked traits." Wilson had no difficulty extending the idea of X-linked traits to humans and declared red–green color deficiency in humans as an X-linked trait.

The discovery of eosin eye color led to two additional changes in the concept of mutation. Eosin arose from a stock of white-eyed flies. When mapped, it turned out to be at the same site as white eyes. This clinched the case against the universality of Bateson's proposal that a recessive trait such as albinism was the absence of the unit character for pigment. Bateson had called this the "presence and absence" model of mutation.[8] Most mutations were simple losses of the unit character for a trait. But eosin arose from white, and thus it could not have arisen from a loss of that factor. There were two ways the fly lab tried to explain the origin of eosin. The first was by invoking "reverse mutation." One could argue that there was a "partial reversion" of white to eosin, or one could argue that the normal allele of the white-eyed factor was actually composed of two components, a white factor and an eosin factor. The second explanation suggested to Morgan that if the normal condition or red eyes was also compound and if it were crossed to white, the heterozygotes would yield rare eosins and the inferred other color factor, which might itself be a "simple white."[9] A sizable number of flies were studied, mostly using heterozygous white females, and no such eosins or other color factors of the eye were obtained. Eosin was then interpreted as a "point mutation" that occurred in the white factor, which itself was some sort of point mutation that arose in the normal gene determining red eyes. The working assumption in the reverse mutation model of white and eosin is the partial undoing of the damage by some event occurring near the original one that created the white mutation.

There was a second feature of eosin that was unusual. Most of the X-linked recessive traits that arose, such as vermilion eye color, yellow body color, forked bristles, or rudimentary wings, looked identical in hemizygous males [XY] and homozygous females [XX]. The Y chromosome played no significant role because Bridges discovered that the loss of the Y chromosome in a sperm led to male offspring that were sterile and that expressed the maternal X. Bridges had worked out not only losses of sex chromosomes but also additions of sex chromosomes.[10] Females having XXY chromosome composition were indistinguishable from females that had XX chromosomes, and they were fertile. The Y played no physiological role in the expression of genes in the females for the eye colors that Bridges followed. He called the new phenomenon "non-disjunction" and identified it as an error of cell division, especially during sperm or egg production.

That meant that the dosage difference between the sexes—two factors or genes in the female and one gene in the male for a given sex-linked trait—was compensated by some mechanism. The phenomenon of darker-eyed females than males in a stock of eosin was described at first as "bicolorism."[11] Because other eye color mutations did not show bicolorism, the normal situation involved a special mechanism that somehow equalized the dose differences, or the threshold for expression in most traits was reached by the single dose itself. It would take another 15 years before an explanation for this bicolorism would be put to an experimental test. It would take 40 years before the phenomenon was described in humans and a different mechanism and terminology were invoked to explain it. Sometimes concepts and hypotheses arise earlier than the tools that are available to explore them. That was the case for dosage compensation by modifying genes in fruit flies and for dosage compensation by X-inactivation in mammals.

The Fluctuating Variation Idea Is Transformed

When Morgan found a mutation, he or one of his students would attempt to generate a stock from it. Many variations were easy to describe and establish as stocks, not only among X-linked traits but among autosomal recessive and dominant traits as well. Flies with curly wings, a second chromosome dominant mutation, showed such stability, although the trait seemed tied to a repressor of crossing over. Recessive traits like vestigial wings, brown eyes, scarlet eyes, and black body color were also fairly consistent in their expression and ease of maintenance. Two dominant mutations that Morgan found, however, failed to become homozygous, and they were variable in expression. Both of these Muller chose to follow. One was beaded wings, which showed a scalloping of the edges of the wings. The other was truncate wings, which showed a lopped off or blunt appearance at the tip of the wings. It took several years for Muller to work out the genetic basis of these two traits, and for the truncate analysis he worked with Edgar Altenburg (1888–1967).[12]

In the beaded case, the failure to establish homozygosity was identified as a dual characteristic of the gene mutation: It was homozygous recessive for a lethal factor that killed it in early embryonic development, but it was dominant in the heterozygous condition and expressed the beaded wing. Later another recessive lethal arose in a third chromosome near the normal allele of beaded, and this permitted a stock of beaded wings to perpetuate itself as a hybrid. The chromosome carrying beaded died when homozygous, and the chromosome bearing the newly arising nearby lethal also died when homozygous. Hence only heterozygotes survived. Muller called these "balanced lethals," and he applied the concept in several ways. He interpreted some of de Vries's new species as examples of balanced lethals, which was later confirmed. He also claimed that the sudden appearance of numerous changes in an individual offspring from such a balanced lethal in the primroses was due to crossing over that allowed accumulated recessive mutations to be expressed in the homozygous condition. That too was later confirmed in *Oenothera*. Muller attributed the variability in expression of beaded progeny to "modifier genes." In the truncate situation, Muller was able to isolate each modifier gene and designate it as an "intensifier" or "diminisher" of the trait expressed by the "chief gene."

Muller also noted that temperature affected gene expression; high temperatures showed the mutant expression more intensely, and lower temperatures tended to normalize the expression of the chief gene. Muller could combine several modifying genes and predict, at a given temperature, the percentages of the different gradations of beaded or truncate wings that would be obtained.

Muller looked on genetic analysis as a process similar to dissection in anatomy and thought that by careful design and application he could isolate the factors involved in a complex variable trait. He rejected the idea that the genes themselves were fluctuating.

In an earlier dispute with William Castle in 1914, Muller had argued that the variations that appeared in a recessive trait like spotting in rabbit fur or hoodedness (a recessive trait in rats) were not, as Castle claimed, fluctuations of the gene for spotting or the gene for hoodedness.[13] Instead, Muller claimed that it was "residual inheritance" that provided the variation. At that time, Muller was just starting his work on beaded and truncate, and it would be several years before he had convincing evidence for the modifier genes and their effects. Instead, Muller relied on the earlier work (appearing in 1903–1909) of Wilhelm L. Johannsen, who showed that variation in bean size was largely a consequence of a large number of factors present in the outbred stock.[14] When the bean plants were highly inbred and intensely selected for small bean size, the variations diminished and eventually a "pure line" would be established in which the range of variation was constant and associated with the environment, the residual inheritance in the inbred line being homozygous. It was this work that led Johannsen to coin the terms "genotype" (the inferred constitution of the individual) and "phenotype" (the appearance of the trait in a given individual).[15] Castle rejected Muller's arguments from Johannsen's work, but he eventually used Muller's findings in beaded wings and agreed that modifier genes were the cause of the fluctuations he reported.

The fly lab worked on numerous projects, collectively and individually. They all noted that most recessive genes kept heterozygous over many generations do not lose their characteristic phenotype when rendered homozygous.[16] This was an issue in the first two decades of the 20th century because many biologists believed that heterozygosity led to "genic contamination." Morgan even believed this in the early days of his genetic work before he discovered sex-limited traits and crossing over. Johannsen's work did not really help explain the fluctuations in phenotype seen in traits like beaded or truncate wing, where the variations were noticeable in each generation. The idea of modifier genes suggested by Muller was at first theoretical, and Castle invoked an appeal to Occam's razor, arguing that modifier genes were ad hoc assumptions lacking experimental evidence. To Castle, the varying phenotype was a reflection of a varying gene that was giving off Darwinian fluctuations.

The fly lab recognized not only the legitimacy of chief genes and modifiers as playing a role in variable traits, but also in providing the genetic basis for Darwinian evolution by natural selection of such small variations. In Muller's mind, it was extended to the evolution of dominance, the new trait being stabilized by the presence of eventually homozygous modifiers. More interesting was the extension of genetics to the interplay of genes in bringing about a phenotype. This worked in two ways. The fly lab noted that many recessive as well as dominant traits had effects on several organ or tissue systems in the fly. A mutation like *dumpy* could affect wing shape, bristle distribution, and the association of the cuticle to the underlying living tissue of the thorax, producing pits, creases, or mounds. Single genes could thus affect many organs, and this was called "pleiotropism." At the same time, there were numerous eye color mutations, and these fell into three categories. One was a loss of a brown pigment, producing bright orange eye colors like vermilion and scarlet. Another was a loss of the bright orange pigment, producing eye colors like brown or sepia. In addition to these two pigments, there was a gene associated with the deposition of the pigments in the individual eyelets. The white-eyed series of mutations with names like *apricot, eosin, cherry, buff,* or *ivory* added as a superscript to the symbol *w* established a "multiple allelic series" (e.g., w^a, w^e, w^{ch}, w^e, w^i, w^+).[17]

The mutation process was not an all-or-none phenomenon. Nor were gene mutations some sort of agents that individually could produce an organ. The gene whose mutation resulted in white eyes did not produce a red eye. Red eye color was the product of an estimated 50 genes that affected the production of the two pigments and the structural elements of the eyelets that were associated with storing and distributing these pigments. Mutations could have reduced activity like yellow or tan body color, and some showed increased activity, such as the body color mutations *sooty*, *black*, and *ebony*. Sometimes that range was characteristic of a multiple allelic series, and sometimes it was apparent by comparing all the mutant genes that were associated with a specific trait like body color. The recognition that eye color in fruit flies involved two different pigments came from the genetic construction of stocks like *bw/bw; st/st*, in which the flies homozygous for both brown eyes and scarlet eyes were white-eyed and indistinguishable in phenotype from the X-linked white-eyed mutation.

Bridges, who became the stockkeeper for the fly lab, collected much of this information on the characteristics of individual mutations. Each of the new mutations was mapped, and a careful drawing of the trait was made (in the first 40 years of the fly lab's existence by Edith Wallace [1881–1964]). Each member of the fly lab supplied this information to Bridges, and his notes found their way into print. The first was through the Carnegie Institution of Washington's publications on the life sciences, a valuable source of detailed knowledge for laboratories working in the field. In the 1930s, Bridges and his coauthor Katherine Brehme (1909–1991) put the notes and illustrations into a reference volume known to a generation or more of fruit fly geneticists as "Bridges and Brehme."[18]

The analysis of the stages at which mutations arose was also significant in characterizing the mutation process. In Muller's analysis of the white-eyed series of alleles, he demonstrated that some arose as point mutations in individual eggs. Some arose earlier during egg formation, and this would lead to a few male offspring with a new *white* allele and a far larger number of male offspring with the normal red eye color. These could be of oogonial origin (when the eggs were being proliferated by mitosis before entering meiosis). They could also represent a mutation that was present in some of the body tissue but not the head region. Sometimes the mutation arose after fertilization, resulting in an individual that is visibly mosaic. These mosaic mutations more often than not did not transmit the mutant trait to their progeny. They were classified as "somatic mutations" in contrast to "germ-line or germinal mutations." If all the offspring showed the mutation in the body tissue and the germinal tissue, this mutation was called a "complete mutation." Muller concluded that both sperm formation and egg formation could give rise to new mutations and that the majority of them were associated with events during meiosis or shortly before or shortly after meiosis, which Muller referred to as a "perifertilization" period of the fly's life cycle.[19]

Mosaics were useful because they also illustrated that most point mutations were "autonomous," and in a typical mosaic, one eye might be red and the other white in a male with a newly arising mutation. Sturtevant recognized that this was not universal.[20] The mutation *vermilion* did not express if it were on a chromosome with other genes serving as markers of its presence. Thus, an X chromosome with yellow body color, vermilion eyes, and forked bristles might be heterozygous in a female that undergoes an early somatic nondisjunction, resulting in a portion of her cells being hemizygous for *yellow vermilion forked*. The resulting mosaic, however, looks yellow and forked along one side of the body axis, and the normal heterozygous tissue occupies the rest of the body. Such a mosaic reveals that the *vermilion* mutation is compensated physiologically by something diffusing from the tissues that are heterozygous. The *vermilion* is then defined as "nonautonomous."

Note that all of these insights into the mutation process were observed and published by 1920. It would be seven more years before mutations were induced by X rays. As each new attribute was teased out by observation and experiment, the sophistication of the gene and its relation to the phenotypes it controlled increased. In just one generation, the sports, atavisms, rogues, plasticity, and polygon of instabilities had disappeared in discussions of mutations. Aboriginal laws had yielded to the chromosome theory of heredity and the theory of the gene.

Not All Mutations Are Autonomous Point Mutations

Sabra Colby Tice (1885–1971) in Morgan's laboratory discovered the X-linked mutation *bar eyes* in 1913.[21] It was a dominant mutation that showed a narrow bar instead of a round compound eye in the male or the homozygous female. But in the heterozygous female, *bar* was expressed as a somewhat broader, kidney-shaped eye. After the stock of bar eyes was established, Morgan and Sturtevant noted that on occasion *bar* reverted to normal and a male with round eyes or a female with kidney-shaped eyes would arise. It was almost always associated with a single event, unlike the white-eyed mutations that seemed to arise in perifertilization stages of the life cycle. This suggested to Morgan and Sturtevant that the event might be associated with meiosis itself, and they placed marker genes on each side of *bar* and looked for these solitary events. They found that they did, indeed, arise in meiosis and were always associated with an event during crossing over because the outside markers (in either direction) would show that a crossover event had occurred. This suggested that *bar* was actually some sort of duplicated gene and during alignment it could lead to a crossover, producing a return to the single-gene state or a triplication resulting in a more extreme eye reduction than *bar* itself in the male. They called the process "unequal crossing over" and attributed it to the random pairing of the two duplicated genes in each of the two chromosomes.[22] Thus, if one chromosome has elements 1 and 2 and if the homologous X has elements 3 and 4, with all four of the elements being identical, the pairing of 1 and 3 or of 2 and 4 leads to no unequal crossover event and only bar-eyed progeny arise. But if 1 pairs with 4 or if 2 pairs with 3, an unequal crossover event will yield a normal eye and a triplicated gene expressing the more extreme bar eye phenotype. There was clearly no point mutation involved, and the mutant expression was assigned to what Sturtevant and Morgan called a "position effect." This implied that most genes were autonomous in expression, some genes could be nonautonomous in expression, and some genes could be subject to positional influences that shifted them from a normal to a mutant activity.

Later, when rearrangements could be induced by mutations, it was noted that position effect mutations were sometimes induced near the site of breakage and that the gene involved could "revert" back to normal if a crossover occurred between it and the breakpoint of the rearrangement. Nikolay Petrovich Dubinin (1907–1998), among others, showed this for bar eyes.[23] Also, Jack Schultz (1904–1971) and other investigators noted that X rays sometimes induced a condition of variable expression of a mutant autonomous gene.[24] If the break was near *white*, the *red* eye allele would show streaks or patches of white eyelets and so would the subsequent generations of offspring receiving that chromosome. The fly lab referred to these as "ever-sporting displacements." Even more remarkable was Schultz's finding that variegated mutations of this sort when supplied with an extra Y chromosome would be rendered normal. Thus, it was not position alone that induced the variegated expression of a nearby gene but some sort of physiological change that could be compensated by the heterochromatin in the Y chromosome (or by additional heterochromatin inserted into the X chromosome).

Chromosome Rearrangements Complicate the Interpretation of a Mutation

The terminology and the complications of describing new findings in a rapidly expanding field tend to confuse both students and practitioners. The situation becomes even more difficult when much of the initial terminology is borrowed from prior work with other organisms and with mental habits of an earlier generation. As the fly work increased and as geneticists sought to expand their knowledge by studying new organisms for their modes of inheritance, the terminology used by different groups could be misleading or confusing. The term mutation was being applied to events that with more careful genetic analysis could be described as polyploidy (the gain or loss of a complete set of chromosomes), aneuploidy (the gain or loss of one chromosome in an otherwise diploid organism), and structural rearrangements. The structural rearrangements included physical changes of the chromosome leading to inversions of segments, duplications of segments, and deletions of segments of a chromosome. They could also involve an exchange of pieces of nonhomologous chromosomes resulting in translocations. Although all of these chromosomal events could be considered of evolutionary interest in some species, as were aneuploids, polyploids, and structural rearrangements in the evolution of *Oenothera*, they were neither the Darwinian variations nor the "point mutations" of the fly lab.

Muller argued that the term "mutation" should be restricted to "variation due to change in the individual gene."[25] The gene was also seen by Muller as the source of the other components in the cell, and the genes determined the life cycle of the organism. It also was the gene that was associated with the evolution of life. It was the source of modifier genes and chief genes, and it was the ultimate progenitor of all genes and hence all life on earth once the first replicating ("autocatalytic") gene arose in earth's evolutionary past. Hence Muller equated the point mutation gene with "the gene as the basis of life."[26] He also predicted that it shared with the newly discovered viruses a capacity for replicating and carrying out a life cycle, and thus genes could become accessible through a chemical and physical study of viruses as possible "naked genes."[27]

References and Notes

1. Morgan TH. 1910. Hybridization in a mutating period in *Drosophila*. *Proc Soc Exp Biol Med* **7:** 160–161.
2. Morgan TH. 1910. The method of inheritance of two sex limited characters in the same animal. *Proc Soc Exp Biol Med* **8:** 17–19.
3. See Chapter 7, "Crossing over versus repulsion," in Carlson EA. 1966. *The gene: A critical history*. Saunders, Philadelphia; and Chapter 13 in Cock AG, Forsdyke DR. 2008. *Treasure your exceptions: The life and science of William Bateson*. Springer, New York.
4. Morgan TH. 1911. Random segregation versus coupling in Mendelian inheritance. *Science* **34:** 384. doi: 10.1126/science.34.873.384.

5. Janssens FA. 1909. La théorie de la chiasmatypie: Nouvelle interprétation des cinèses en maturation. *Cellule* **25:** 387–411.
6. See Carlson EA. 2004. *Mendel's legacy: The origin of classical genetics,* p. 310. Cold Spring Harbor Laboratory Press, Cold Spring Harbor, NY.
7. Sturtevant AH. 1913. The linear arrangement of six sex-linked factors in *Drosophila,* as shown by their mode of association. *J Exp Zool* **14:** 43–59.
8. See Chapter 8, "The presence and absence hypothesis," in Carlson 1966 (note 3).
9. Carlson 1966 (note 3), pp. 194–196. Hyde RR. 1920. Segregation and recombination of the genes for tinged, blood, buff, and coral in *Drosophila melanogaster. Proc Indiana Acad Sci* **1920:** 291–300.
10. Bridges CB. 1916. Non-disjunction as proof of the chromosome theory of heredity. *Genetics* **1:** 1–52; 107–163.
11. Morgan TH, Bridges CB. 1913. Dilution effects and bicolorism in certain eye colors of *Drosophila. J Exp Zool* **15:** 429–466.
12. Muller HJ. 1917. An *Oenothera*-like case in *Drosophila. Proc Natl Acad Sci* **3:** 619–626. Muller HJ. 1918. Genetic variability, twin hybrids, in a case of balanced lethal factors. *Genetics* **3:** 422–499. Altenburg E, Muller HJ. 1920. The genetic basis of Truncate wing—An inconstant and modifiable character. *Genetics* **5:** 1–59.
13. Muller HJ. 1914. The bearing of the experiments of Castle and Phillips on the variability of genes. *Am Nat* **48:** 567–576.
14. Johannsen W. 1909. *Elemente der exakten Erblichkeitslehre.* G. Fischer, Jena, Germany.
15. Ibid. (An English translation of that definition of the gene is in Carlson 1966 [note 3], p. 20.)
16. Muller HJ. 1916. The mechanism of crossing over I–IV. *Am Nat* **50:** 193–221; 284–305; 350–366; 421–434.
17. Morgan TH, Bridges CB. 1919. Contributions to the genetics of *Drosophila melanogaster.* I. The origin of gynandromorphs, Publication No. 278, pp. 1–172. Carnegie Institution of Washington, Washington, DC. Bridges CB, Morgan TH. 1919. The second chromosome group of mutant characters of *Drosophila melanogaster,* Publication No. 278, pp. 123–304. Carnegie Institution of Washington, Washington, DC.
18. Bridges CB, Brehme K. 1944. The mutants of *Drosophila melanogaster,* Publication No. 552. Carnegie Institution of Washington, Washington, DC. Bridges died in 1938. He had submitted a first draft that was published in an annual newsletter, *Drosophila Information Service,* Number 9. Brehme updated the mutants to 1944, and Carnegie published the compendium as a paperback.
19. Muller HJ. 1920. Further changes in the white-eye series of *Drosophila* and their bearing on the manner of occurrence of mutations. *J Exp Zool* **31:** 443–473.
20. Sturtevant AH. 1920. The *vermilion* gene and gynandromorphs. *Proc Soc Exp Biol Med* **17:** 70–71.
21. Tice SC. 1914. A new sex-linked character in *Drosophila. Biol Bull* **26:** 28–30.

22. Sturtevant AH, Morgan TH. 1923. Reverse mutation of the *bar* gene correlated with crossing over. *Science* **57:** 746–747.
23. Dubinin NP, Sidorov BN. 1934. Relation between the effect of a gene and its position in the system. *Am Nat* **68:** 377–381.
24. Schultz J. 1936. Variegation in *Drosophila* and the inert chromosome regions. *Proc Natl Acad Sci* **22:** 27–33.
25. Muller HJ. 1922. Variation due to change in the individual gene. *Am Nat* **56:** 32–50.
26. Muller HJ. 1926. The gene as the basis of life. *Proc Fourth Int Congr Plant Sci (Ithaca)* **1:** 897–921.
27. Op. cit., note 25, p. 50.

5

Radiation and the Analysis of Mutation by Mutagenesis

THOMAS HUNT MORGAN, FERNANDUS PAYNE, James W. Mavor, and Albert F. Blakeslee tried to induce mutations with X rays before Hermann Joseph Muller's publication of 1927, but they failed.[1] What they shared in common was a lack of knowledge about spontaneous mutation rates and, by default, an absence of knowledge of what to look for other than departures from normalcy in appearance. They also had no idea what dose of radiation would be effective in an era that lacked dosimeters and that tested radiation effects by sticking one's hand under an X-ray machine and exposing it until the skin reddened. Doses below this physiological response were considered by health practitioners to be below a "threshold dose" and thus harmless to biological processes. X rays had been found by Wilhelm Roentgen (1845–1923) in 1895 and rapidly entered the life sciences. By 1900, many physicists and physicians reported biological effects of radiation, including reddening of the skin, slow-healing blisters or ulcerations of the skin, numbness of the fingers used to hold photographic plates while administering X rays, and cancers, many of them fatal. French investigators also found that X rays had effects in producing sterility, causing aborted embryos in frogs, or even led to parthenogenetic development if the sperm were heavily irradiated. They also noted a lot of fragmentation of the chromosomes.[2] No doubt reports of these adverse effects of radiation stimulated interest by Morgan and others to try X rays.

From the start of the Columbia fly lab to 1926, other agents were tried as possible agents that could induce mutations, including ether and alcohol. Muller took an interest in measuring mutation rates and inducing mutations when he moved to Rice University. With Edgar Altenburg, he obtained the first measure of spontaneous mutations in fruit flies using a category that

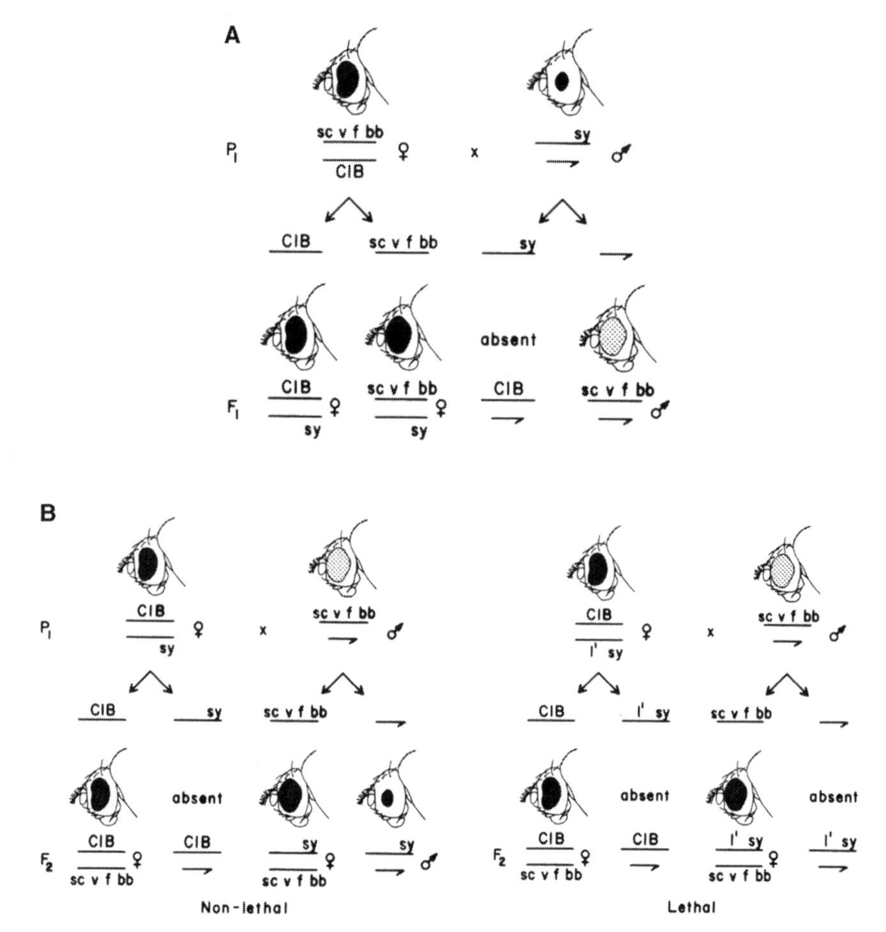

The ClB technique establishes a quantitative and objective method for obtaining mutation rates. H.J. Muller designed the ClB stock to make the detection of X-linked recessive lethal mutations easier to find. He later replaced it with a crossover suppressor that lacked a recessive lethal of its own and which retained Bar eyes as a dominant mutation. It also has a recessive gene, apricot eye color. It is called the Basc *chromosome, and it is widely used as a balancer to keep stocks of X-linked mutants that are infertile, frail, or lethal. It is also used for studies involving X-linked mutation frequencies. (Reprinted from Carlson EA. 1981.* Genes, radiation, and society: The life and work of H.J. Muller. *Cornell University Press, Ithaca, NY.)*

was fairly abundant. Morgan called these "lethal mutations." He had found the first such case in 1912 when it appeared as an altered sex ratio, half the sons being absent.[3] Muller and Altenburg decided to test females individually in vials and see how many showed a loss of half their sons. They found a mutation rate of about one sex-linked lethal per hundred X chromosomes,[4] but later experiments attempting to identify mutations by using a stock that made it easier to reveal X-linked lethals showed that the rate was variable,[5] and the figure Muller obtained was now one X-linked lethal per thousand sperm tested. The stock Muller designed, called *ClB,* used an X chromosome with the dominant marker bar eyes (the *B* of the *ClB*) and a "crossover suppressor" (the *C* of the *ClB,* later identified as an inversion) and a recessive lethal (the *l* of the *ClB*). That chromosome could serve as a balancer for induced or spontaneously arising X-linked lethal mutations. Any such female receiving a sperm with an induced X-linked lethal would produce only daughters because either of the X chromosomes if fertilized by a Y-bearing sperm would abort development of the son.

Muller thought that the variable rate of spontaneous mutation might be due to "mutator genes" that somehow altered the physiological basis for mutation production. In the early 1920s, Muller considered mutation to be primarily a chemical process and thus he used temperature difference (10°C) as one way to look for a higher mutation rate. While doing these experiments, he noted that occasionally, at the same temperature, the mutation rate could differ. Whether these differences were caused by his inferred mutator genes or if they reflected, instead, some incompatibility associated with transposable elements or some infective agent is not possible to determine because the strains these investigators used are no longer in existence or may have undergone their own physiological changes over the years.

While on sabbatical leave at the University of Chicago, Muller read the literature on the biological effects of radiation and revived that possibility. He concluded that X rays would act as a "punctiform agent," with each encounter of a gene and X ray localized in its effect. He was now equipped with the following advantages that were impossible for Morgan and others about 1910. He knew that the spontaneous mutation rate would be about one X-linked lethal per thousand. He would be looking not for vague departures from normalcy but for a specific type of point mutation that killed the sons if their father's sperm carried it. He also used a very high dose (estimated today at about 4000 roentgens) after consulting with a physicist. The experiments began in the fall of 1926 and were completed in early 1927. The results were startling. Instead of a doubling of the mutation rate, Muller was obtaining a rate about 150 times the spontaneous frequency. In addition to the expected X-linked lethals, he found numerous X-linked visible mutations characteristic of those the fly lab had found as spontaneous mutations over the previous 15 years. They were also as numerous as those found over that time period, and thus X rays could be used as a tool to produce visible mutations and saturate the X chromosome with its possible visible variants. Most turned out to be point mutations that Muller could map. He tested the familiar visible mutations with the spontaneous form, and most showed that they were allelic. There were also several lethals that could not be mapped. Muller put these aside, but when he spent a summer in New York, using the facilities at New York University, he lost all those stocks because a person charged with cleaning glassware in the laboratory thought they were dirty glassware intended for cleaning.

Muller published his first results in *Science* in 1927 but did not include his data.[6] He just wanted to establish his priority for a major discovery. His data followed in the July 1927 meetings of the International Congress of Genetics in Berlin.[7] Until those data appeared, Morgan believed that Muller had committed academic suicide because he had already mentioned to other geneticists that the mutation rate he reported in 1919 was not confirmed by his later tests using his new stock design. The inconsistency of Muller's work disturbed Morgan. What Morgan could not have known at the time was the variability of the mutation rate, which is dependent on a background genotype, and the infectious nature of some transposable elements that can disrupt the genome, leading to higher mutation rates.

Muller also reported another class of mutations he found in his first X-ray experiments. Many of the fertilized eggs aborted and did not hatch as larvae. Muller called them "dominant lethals." At the time, the mechanism of their formation was unknown. Some thought they were point mutations that, like recessive lethals, were fatal to development, but as dominant mutations they would kill in a single dose. This interpretation was offset by their numerical abundance. There were more such dead embryos than all of the recessive lethals obtained. It did not seem likely that dominant lethals would behave differently from visible dominant point mutations, which were rare (one-tenth as frequent) compared to recessive visible mutations.

Muller's work was rapidly confirmed by the publication of Lewis Stadler's experiments on mutations in barley and other plants.[8] He induced visible mutations with X rays, but by that time plant biologists, including Stadler, using dozens of species, had reported newly arising variants involving plant organs and their properties. Stadler was familiar with the visible variations that might arise, a limitation that Muller avoided by using a quantitative approach in which vials were scored as having or not having any sons in the F_2 of the ClB crosses. Any fruit fly investigator did not require years of experience learning and describing mutant phenotypes. Soon dozens of investigators were using X rays and confirming Muller's results. The field of radiation genetics was rapidly expanding.

Among the findings by Muller and these early investigators was the relation of point mutations to dose. They were linear. This occurred whether the total dose was administered in a short time (e.g., 30 minutes) or over a long time (e.g., 30 days).[9] The amount of time, however, was significant for the production of chromosomal rearrangements, such as inversions or translocations. These were exponential with dose, somewhat less than the theoretical square of the dose difference (e.g., a twofold increase in dose would lead to a fourfold increase in translocations).[10] That lower result (about to the 3/2 power instead of the square) was attributed to the facts that radiation was administered to mature sperm in fruit flies and the chromosomes were compactly arranged, allowing individual paths of X rays to pass through multiple overlapping chromosomes. If the dose were spread over a long time (days or weeks instead of minutes), the percentage of translocations or inversions but not point mutations would diminish. Later Muller would interpret this drop in translocations in protracted exposures to a dose of radiation as an outcome of the loss of those chromosomes with single breaks that could not find another break elsewhere in the nucleus. They would end up as "dominant lethals."

A dispute arose over the nature of mutations induced by X rays. They were mostly point mutations, but were they the same as point mutations arising spontaneously? Stadler believed that most, if not all, of his X ray–induced mutations were minute rearrangements within a gene. Such "intragenic minute rearrangements" would be consistent with a view that X rays are limited to breaking chromosomes.[11] It would be another generation before geneticists could discuss the mutation process with X rays in terms of nucleotide changes and sequences where such disputes could be put to a test. But even before that molecular era came into being, experiments in the 1940s by Wilson S. Stone (1907–1968), Orville Wyss (1912–1993), and Felix Haas (1917–) demonstrated that there was an "indirect effect" of radiation.[12] It was known for some time from studies by chemists and physicists that X rays alter water, leading to the formation of hydrogen peroxide, reactive molecules, and ionization, all of which promote chemical reactions in the vicinity of the genes. Stone and colleagues showed that if Petri dishes bearing agar food media are irradiated with X rays and the plates are then streaked with bacteria, the bacteria produce about the same number of mutations as bacteria irradiated directly on the plate. Control plates streaked with nonexposed bacteria produced few mutations. This suggested that the major cause of mutations in a water-based medium was chemical. Hydrogen peroxide was by this time known to be a chemical mutagen.

Muller and other fruit fly geneticists developed stocks to study breakage events. A microscope is not needed to detect translocations or inversions. Translocations can be detected by a genetic method in which two nonhomologous chromosomes are marked with mutations that assort independently (yielding a 9:3:3:1 $F_1 \times F_1$ ratio or a 1:1:1:1 test cross-ratio). Using the eye-color markers *brown* (*bw*) and *scarlet* (*st*), irradiated sperm that are normal for both color factors fertilize nonirradiated eggs that are *bw;st*. This produces F_1 flies that are red-eyed and dihybrid *bw/+; st/+*. If those dihybrid males are individually mated to homozygous *bw/bw; st/st* females, there should be white (*bw;st*) brown (*bw;+*), scarlet (*+;st*), and white (*bw;st*) flies in a vial among the F_2 generation. But if a translocation connects a region of chromosome II bearing the normal allele of *brown* with a region of chromosome III bearing the normal allele of *scarlet*, the resulting F_2 offspring will only produce red-eyed flies (with the translocation heterozygous for the recessives *bw* and *st*) and white-eyed flies (with both *bw* and *st* in the homozygous condition). The other two categories fail to develop because of the aneuploidy associated with the translocated component and normal chromosome components going into a sperm. What makes genetics attractive to scientists is the versatility of designing genetic stocks that can solve problems. Muller was a virtuoso in such stock design, and many of his stocks for detecting X-linked lethals, for detecting translocations, for maintaining lethal or sterile recessive mutations using a balanced lethal design, or even for determining the size or boundaries of genes are very sophisticated. These stock designs were intimidating to graduate students until they learned to use them in their experiments.

In 1916, when Morgan and his students had amassed an abundance of evidence for the chromosome theory of heredity first proposed by Walter Sutton, they could show a map of the four chromosomes of the haploid and designate the chromosomes as X, II, III, and IV.[13] All of the point mutations mapping to the X chromosome showed Morgan's sex-limited inheritance. The genes mapped to chromosomes II and III revealed chromosomes that were about twice the length of the X chromosomes, and their behavior was characteristically Mendelian for any gene of group II with any gene of group III. The genes on chromosome IV were sparse, and cytologically this chromosome was the smallest of the four chromosomes. The only fourth chromosome gene then known (*bent wings,* found by Muller) would show a Mendelian ratio with any member of groups II or III. The Y chromosome had been added a year earlier when a rereading of Nettie Stevens's papers confirmed that she had first found the Y chromosome in fruit fly males and Morgan had misinterpreted her figures as showing no independent Y chromosome for fruit flies. No Y chromosome–associated genes were reported in that first book-length treatment of fruit fly genetics.

By the time Calvin Bridges and Katherine Brehme published their book on the mutations of *Drosophila melanogaster,* there were hundreds of mutations that had been described and mapped.[14] Genes on the Y chromosome remained rare, but this was consistent with the idea that none of its genes were essential for the life of the fly because XO males, other than being sterile, were as robust as XY males and sexually active. But during that 20-year stretch between *The Mechanism of Mendelian Heredity* and *The Mutants of Drosophila melanogaster,* a lot had been learned about the relation of mutations to their cytology. Some regions of the chromosome contained very few genes. These were referred to as "inert regions" of the chromosome, and most of these turned out to be in material called "heterochromatin," which could be stained differentially. The more compacted chromosome regions associated with the bulk of point mutations were found in the "euchromatin," and they took on a darker stain. Acetocarmine was the dye of choice for these studies.

Heterochromatin also was associated with numerous rearrangements from breakages induced by X rays. This reflected the paucity of genes in this region. A major shift in the relation of mutations to their chromosomal location came from the discovery by Theophilus S. Painter (1889–1969) in 1934 that there were giant chromosomes in the salivary glands of fruit fly larvae.[15] These turned out to be cable-like arrangements of replicated chromatids that remained partially unwound. Very quickly Painter and his Texas colleagues, Muller (then in the USSR) and his colleagues, and Bridges (at The California Institute of Technology, Caltech) were in a race to examine giant chromosomes for a variety of genetic issues. All three noted that the organization of these giant chromosomes produced a reliable banding pattern that they could examine and relate to mutations associated with position effects, deletions, translocations, or other chromosomal events. The more compact regions were called "bands" and assigned to the euchromatin.

The nonbanded, paler areas were assigned to "interstitial heterochromatin" because they were finely dispersed between inferred genes or clusters of genes. The heterochromatic regions seen in mitotic chromosomes were much smaller in the giant chromosomes, but as in the mitotic chromosomes, there was a preponderance of that heterochromatin associated with the tips of chromosomes and the region around the centromere. In the spreads produced by bursting the nuclei containing giant chromosomes, there were five noticeable arms, one representing the X chromosome, two representing chromosome II, and two representing chromosome III. A small nubbin represented chromosome IV. All of these arms and the nubbin were attached to a central region forming the heterochromatic regions of these chromosomes.

Muller's experiment also ruled out a model of the gene composed of numerous particles. That "genomere" model assumed that it would explain the fluctuating variations associated with some variable traits.[16] When Muller tested the chromosomes of nonlethals into the F_2, F_3, or F_4 generations, he found no such occurrence after the F_2 generation. There were a small number of mosaic X-linked lethals in that generation. Muller interpreted them then as arising from prematurely doubled chromatids in the sperm (i.e., they were chemically doubled but visibly seen as single chromatids in the mature sperm). Almost 30 years later, that "premature doubling" was shifted from the chromatid to the structure of DNA itself as a double helix. A few visible mutations were also fractional in their origin from X-rayed sperm. Whatever the mechanism was for inducing mosaic origins of point mutations, they were in the minority of all X ray–induced point mutations. This supported Muller's working hypothesis when he chose to use X rays to induce mutations—that the ionizing radiation was probably "punctiform" in its mechanism and localized to regions within the individual genes.

Radiation Can Be Used as a Tool to Construct Stocks to Study the Properties of Mutations

Muller clarified the types of structural mutations produced by radiation. He reported the presence of translocations, inversions, deletions, and duplications. Some of the rearrangements involved minute segments of the chromosome. Sometimes a gene and some of its immediate neighbors were shifted from one chromosome to another. This happened with the X-linked trait called *sc*^{19} (*scute-nineteen*). It was transported and inserted into the second chromosome near the *dumpy* locus (at 13.0 on chromosome II).[17]

It functioned in this new environment with some position-effect irregularities of the scute bristles, but it also carried the normal allele of yellow body color and achaete bristles. This allowed Muller to construct stocks that contained yellow on the X chromosome but expressed the normal amber color because the normal allele for yellow was inserted on chromosome II. He also constructed stocks to test for the presence of a class of modifier genes on the X chromosome that bring about "dosage compensation."[18] He could use X rays to trim chromosomes so that the gene region he desired was capped by the heterochromatic tip of the largely euchromatic arm of the X chromosome, and at the other end, it would be capped by the heterochromatic region of the other arm with its centromere. Such "deleted" chromosomes could test dosage relations of X-linked genes in hemizygous males or homozygous females. Muller demonstrated that bicolorism was actually a consequence of dosage compensation. Curt Stern (1902–1981), using the *bobbed* gene on the X and Y chromosomes, independently confirmed his work; *bobbed* was one of the few genes on the Y chromosome that could be detected by a physical body change, in this case of bristles.[19]

When Muller obtained X ray–induced mutations near the *y ac sc* (yellow body color, achaete bristles, and scute bristles) region at the tip of the X chromosome about a map unit to the left of the white-eyed region, many of them had an inversion break in or near the centromere region on the other end of the X chromosome. Muller used crossing over to combine a piece of one inversion with a reciprocal piece of the other inversion to produce recombined inversions that had duplications of part or all of the *y ac sc* region or deletion of part or all of the *y ac sc* region. Most interesting were those inversions that combined portions of two inversions with no visible change in the *y ac sc* region. Muller called this the "left–right test."[20] It suggested that genes had boundaries and were embedded in material that could break in different but nearby places without producing any change in expression of these three genes.

Bridges at Caltech and Muller in Moscow reported independently the most significant of the findings from salivary chromosome analysis. They had each looked at the *bar eye* gene and found a visible duplication of a small segment of the X chromosome in the salivary gland chromosomes they examined.[21] To Muller this finding provided the mechanism of pairing that resulted in deletions of one segment of the *bar eye* region or a triplication of that segment in the ultrabar or more extreme phenotype that departed from typical bar, which was a duplication. It also provided to Muller a model of the evolutionary origin of new genes.[22] Muller claimed that the original *bar* found by Sabra Tice must have come about from a "primary unequal crossing over," but the presence of the homozygous duplication afforded occasional mispairings or "secondary unequal crossing over" that led to the reversions to round eyes or the presence of a more extreme ultrabar. The duplicated gene in the initial origin of *bar eyes* thus represented a way genes could increase in number by tandem repeats with subsequent differential histories of mutations in the duplicated genes leading to new or cognate functions. What could be called "Muller's doctrine"—that all genes arise from preexisting genes—thus followed, reminiscent of Rudolf Virchow's *omnis cellula e cellula,* except for the first gene, which is associated with the origin of life.

Chemically Induced Mutations Differ in Their Mode of Origin

Although X rays turned out to be potent in their production of mutations, Muller believed that most mutations arising in fruit flies were not a consequence of background radiation. He worked with physicist Lewis Mott-Smith (1873–1952) and obtained the residual background radiation from flies that were roasted to ash; he also obtained the background radiation of the campus building where they were housed. From his assumption of linearity (based primarily on doses ranging from several hundred to several thousand roentgens), he showed that only a minor fraction of spontaneous mutations owed their origin to background radiation.[23] Although Muller suspected that most mutations in fruit flies arose from chemical processes in the cells, he was not able to find suitable chemicals to test for mutation. In 1939, he discussed that possibility with Charlotte Auerbach (1899–1994), and she began a search for chemical mutagens. Muller left for the United States in 1940, and shortly after he left, Auerbach consulted with J.M. Robson (1900–1982), a pharmacologist. He suggested that she try mustard gas because that was a potent agent that produced slow-healing burns similar to those associated with higher doses of exposure to X rays. Auerbach administered the mustard gas, burning her own fingers while administering it, and wrote to Muller that she had found scads of mutations, but she could not divulge the agent she was using because of war secrecy.

Auerbach published her work shortly after the war ended.[24] She found that nitrogen mustard was equally as effective as mustard gas. Both are known chemically as alkylating agents. One surprise was the relative paucity of chromosomal rearrangements at chemical doses that were providing point mutations that were equivalent to X-ray exposures of 4000 roentgens.

A second surprise was the large number of mosaics she obtained, both as X-linked lethals and as visible mutations. Some of these "fractional mutations" Auerbach interpreted as "delayed effects" in which the gene was altered but not yet converted to a mutation. Those in Muller's laboratory concluded, 10 years later, that most of these mosaics were what one would expect if the DNA of the sperm were chemically altered in a single strand of a gene, altering one purine or pyrimidine.

Independently of Auerbach, Joseph A. Rapoport (1912–1990) in the USSR used formaldehyde to induce mutations in fruit flies, but his work was also held up by wartime censorship and not published until after the war ended.[25] Many chemicals were added to the list of potent mutagens. The most widely used for experimental studies was the alkylating agent ethyl methanesulfonate (EMS), which could be fed in a sugar solution instead of being injected. Agents such as quinacrine mustard were also effective in fruit flies but required injection with finely drawn Pasteur pipettes into the abdomens of male fruit flies. A field of comparative mutagenesis emerged in the late 1950s and 1960s, exploring the way different mutagens acted on the DNA or how the mutations differed when induced by different agents. Most of the molecular analysis was done with bacteriophage viruses. Fruit flies were important, however, in interpreting how mutations become fixed and distributed to the somatic tissues and gonadal tissues of the flies.[26] That, in turn, served as a model of the origin and distribution of human mutations that sometimes are expressed as mosaics in their initial appearance. Retinoblastoma was one such mutation of medical interest.[27]

References and Notes

1. Morgan mentions his failure with using radiation in his description of the origin of truncate wings, a mutation he found in 1910. He did not know if he had induced it or if it was one of his "mutating period" mutations. There is no written account of his methods or how intensely he pursued this effort. The account of the discovery of truncate wings is in Bridges CB, Morgan TH. 1919. The second chromosome group of mutant characters of *Drosophila melanogaster*, Publication No. 278. Carnegie Institution of Washington, Washington, DC. Negative results are frequently not published. Muller, in his course on "Mutation and the Gene," and Fernandus Payne, in my interview with him (see Chapter 3, note 38), mentioned Payne's and Albert F. Blakeslee's failure to induce mutations.
2. Bergonie J, Tribondeau L. 1904. Actions des rayons X sur le testicule du rat blanc. *C R Seances Soc Biol Fil* **57**: 400–402.

3. Morgan TH. 1912. The explanation of a new sex ratio in *Drosophila. Science* **36:** 718–719.
4. Muller HJ, Altenburg E. 1919. The rate of change of hereditary factors in *Drosophila. Proc Soc Exp Biol Med* **17:** 10–14.
5. Muller HJ. 1928. The measurement of gene mutation rate in *Drosophila,* its high variability, and its dependence upon temperature. *Genetics* **13:** 279–357.
6. Muller HJ. 1927. Artificial transmutation of the gene. *Science* **66:** 84–87.
7. Muller HJ. 1927. The problem of genic modification. Fifth International Genetics Congress, Berlin, 1927. *Verhandlung des V Internationale Kongress Vererbungslehre. Zeitschrift fur Induktiv und Abstammungs Vererbungslehre,* Supplement 1, pp. 234–260.
8. Stadler L. 1928. Mutations in barley induced by X-rays and radium. *Science* **68:** 186–187.
9. Ray-Chaudhuri SP. 1939. The validity of the Bunsen-Roscoe law in the production of mutations by radiation of extremely low intensity. Proceedings of the Seventh International Congress of Genetics, Edinburgh, 1939 (Supplement), p. 146 [Abstract].
10. Muller HJ, Altenburg E. 1930. The frequency of translocations produced by X-rays in *Drosophila. Genetics* **15:** 283–311.
11. Stadler LJ. 1932. On the genetic nature of induced mutations in plants. *Proc Sixth Int Congr Genet (Ithaca)* **1:** 274–294.
12. Stone WS, Wyss O, Haas F. 1947. The production of mutations in *Staphylococcus aureus* by irradiation of the substrate. *Proc Natl Acad Sci* **33:** 59–66.
13. Morgan TH, Sturtevant AH, Muller HJ, Bridges CB. 1915. *The mechanism of Mendelian heredity.* Henry Holt, New York.
14. Bridges CB, Brehme K. 1944. The mutants of *Drosophila melanogaster,* Publication No. 552. Carnegie Institution of Washington, Washington, DC.
15. Painter TS. 1934. A new method for the study of chromosome aberrations and the plotting of chromosome maps in *Drosophila melanogaster. Genetics* **19:** 448–469.
16. Carlson EA. 1966. The genomere hypothesis. In *The gene: A critical history,* Chapter 12. Saunders, Philadelphia.
17. Muller HJ. 1935. The organization of chromatin deficiencies as minute deletions subject to insertion elsewhere. *Genetica* **17:** 237–252.
18. Muller HJ. 1932. Further studies on the nature and causes of gene mutations. *Proc Sixth Int Congr Genet (Ithaca)* **1:** 213–255.
19. Stern C. 1929. Unterschungen über Aberrationen des Y chromosoms von *Drosophila melanogaster. Z Indukt Abstamm Vererbungsl* **51:** 253–353.
20. Muller HJ. 1956. On the relation of chromosome changes and gene mutations. *Brookhaven Symp Biol* **8:** 126–147.
21. Bridges CB. 1936. The Bar "gene"—A duplication. *Science* **83:** 210–211.

22. Muller HJ. 1936. Bar duplication. *Science* **83:** 528–530. Muller HJ, Prokofeyeva-Belgovskaya AA, Kossikov KV. 1936. Unequal crossing over in the bar mutant as a result of duplication of a minute chromosomal section. *Compt Rend (Dokl) Acad Sci URSS* **1:** 87–88.
23. Muller HJ, Mott-Smith, L. 1930. Evidence that natural radioactivity is inadequate to explain the frequency of "natural" mutations. *Proc Natl Acad Sci* **16:** 277–285.
24. Auerbach C, Robson JM. 1946. Chemical production of mutations. *Nature* **1946:** 157–302.
25. Rapoport JA. 1948. Mutations under the influence of unsaturated aldehydes. *Compt Rend (Dokl) Acad Sci URSS* **61:** 713–715.
26. Southin JL. 1966. An analysis of eight classes of somatic and gonadal mosaicism at the dumpy locus in *Drosophila melanogaster. Mutat Res* **3:** 54–65. Carlson EA, Oster II. 1962. Comparative mutagenesis of the dumpy locus in *Drosophila melanogaster.* II. Mutational mosaics induced without apparent breakage by a monofunctional alkylating agent. *Genetics* **47:** 561–576.
27. Carlson EA, Desnick RJ. 1979. Mutational mosaicism and genetic counseling in retinoblastoma. *Am J Med Genet* **4:** 365–381.

6

Using Biochemical Approaches to Study Mutation

WHEN WILLIAM BATESON FIRST REFLECTED on the unit characters, as he called the bearers of his Mendelian traits, he knew how they were transmitted as Mendelian units with predictable ratios, but he had no clue how they worked, asking, almost plaintively, "What are they?"[1] By 1909, he was more confident that they were basic to all life, Mendelism having been extended to many plants and animals. He also knew that they affected biochemical traits from the work of Archibald Garrod (1857–1936), who consulted Bateson on alkaptonuria, a recessive Mendelian metabolic defect that causes a deficiency in the enzyme that degrades alkapton and thus produces a brownish-black oxidized discoloration of the urine. The high incidence of cousin marriages among the pedigrees of the individuals having this condition suggested this Mendelian interpretation for what Garrod called "inborn errors of metabolism."[2] In the next few years, hemophilia was added as a human disorder that showed X-limited expression. Something in the blood of these males was preventing adequate clotting. Huntington disease was also recognized as a dominant Mendelian late-onset neurological disorder that caused tremors, progressive paralysis, dementia, and death. Such mutations were also being found in laboratory organisms. There were waltzing mice, fruit flies that exhibited tremors, chlorotic or albino plants that failed to carry out photosynthesis, and numerous instances of albinism. Some traits were minor, but many were pathologies that were serious for the development or functioning of a tissue or organ. In each case, the mutation process leading to the recessive or dominant trait altered some normal activity. One could no longer say, as did some biometricians in 1902, that Mendelian traits were trivial traits and had no fundamental effect on the heredity of an organism.

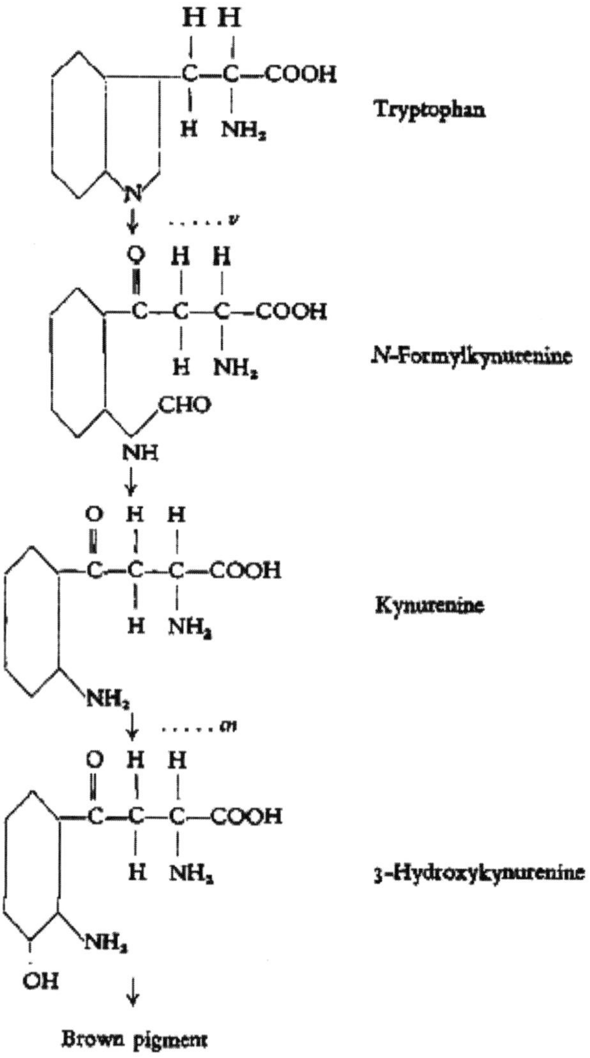

Tryptophan

N-Formylkynurenine

Kynurenine

3-Hydroxykynurenine

Brown pigment

George Beadle and his colleagues work out the biochemical basis of eye color mutations in fruit flies. The developmental approach used by Beadle and Boris Ephrussi required transplants of organ rudiments (anlagen or imaginal discs) that were inserted into larvae of a different eye color. Depending on the outcome of eye color, they could work out a sequence of steps that later turned out to be a conversion of tryptophan by several steps into 3-hydroxykynurenine the brown pigment component of the eye. Note to reader: The somewhat crude look of this illustration is a consequence of its facsimile origin. Unless indicated otherwise, all illustrations in this book are from the original printed sources. (Reprinted, with permission, form Beadle GW. 1958. Genes and chemical reactions in Neurospora. *Nobel lecture December 11. Nobel Award speeches in Medicine and Physiology. © ® The Nobel Foundation.)*

By 1926, when Hermann Joseph Muller proposed his view on "the gene as the basis of life," not much had been added either to the chemical nature of the gene or how genes worked. Most biologists believed that genes were proteins.[3] There was a lot of evidence for this. Even before the discovery of Mendelism, it was known that mammals produced a reaction to the injection into their blood of living matter from other organisms, even from their own species. This led to the discovery of blood groups in 1901 by Karl Landsteiner (1868–1943). The agglutinating components were all shown to be proteins.[4] Also, quite a bit of knowledge was accumulating about proteins. They were known to be large molecules, and it was known that they had enormous diversity because fractionating them from mixed debris of shredded cells produced many antigenic types estimated in the hundreds. Their complexity was partially explained by their use of about 20 different amino acids in their assembly. No one then had a clue as to how to locate the amino acids in a protein or what a protein's shape was.

The other major constituents of cells were carbohydrates, represented by starches or glycogen and composed of sugars. Fats were represented by fatty acids and associated small molecules. These could be found in storage vesicles in fat cells of adipose tissue, or they could be found in the composition of cell membranes. Neither carbohydrates nor lipids were considered to be possible genetic agents. The other contender for universally found components of cells was nucleic acids. Edmund Beecher Wilson considered these as possible agents that could serve as hereditary units, and he wrote about his ideas favoring nucleic acids in 1896 in his book, *The Cell in Development and Inheritance.*[5] He based this conclusion on the high nucleic acid content in sperm and cell nuclei. That view was diminished by the relative paucity of components that composed nucleic acids. It was known there were two purines (adenine and guanine) and two pyrimidines (cytosine and thymine). At Rockefeller Institute, Phoebus Levene (1869–1940) favored a secondary role for the nucleic acids as some sort of scaffolding for the genes. There was little variation he could find in the gross composition of the four nucleotides, which were roughly evenly distributed in animal and plant cells.[6] By default, the proteins were in the lead. They also had known functions, especially as enzymes (originally called "ferments" in the 19th century).

Even with a working model of genes as some sort of proteins or specialized enzymes, there was little insight into how they brought about metabolic activity. Even less was known, save for Bateson's neglected insights into mutations that affected body plans and symmetry, about genes and development. It was widely assumed that genes were silent in most tissues except for those associated with the specific activities of the tissue. From the cytological studies using abundant rearrangements brought about by radiation, Muller was confident that the gene was linear or ribbon-like and not globular, that it had specific boundaries, and that point mutations had the capacity to copy their errors without losing the ability to copy. There was something about position effect that altered gene function without altering the gene the way point mutations did. It was an epigenetic phenomenon. Until the late 1930s, there was no biochemical insight into the functioning of the genes. That insight came as genetics shifted from Muller's approach using radiation and cytogenetics to the work of George Beadle (1903–1989) and Boris Ephrussi (1901–1979), who used fruit flies to establish the first model of gene function.

George Beadle Shifts the Gene to Biochemical Functions

Beadle was an undergraduate from Nebraska with an interest in genetics, and he worked with Rollins Emerson (1873–1947) at Cornell University, taking an interest in maize genetics. He moved on to Caltech, where he studied with Alfred Henry Sturtevant in Thomas Hunt Morgan's laboratory and picked up techniques for fruit fly genetics. Beadle was interested in using a different approach and teamed up with Boris Ephrussi, who was visiting Caltech from Paris. Sturtevant had found some 20 years earlier that vermilion-eyed flies do not show autonomy in mosaics (usually occurring as sex chromosome mitotic nondisjunction expressed in a gynandromorph), but he could not explain why. Something in the heterozygous normal somatic tissue diffused and corrected the functioning of the hemizygous cells in the ommatidia of the eye that was mutant for *vermilion*, and it expressed, instead, red eyes. Beadle and Ephrussi taught themselves how to do a series of transplant experiments in which the anlagen or rudiments (technically known as imaginal discs) that form the future eyes could be inserted from one larva into another of a different genetic strain. They tried placing mutant rudiments into normal flies to see which of the several dozen eye-color mutants were autonomous. Then they conceived of testing pairs of similar mutations such as using scarlet-eyed fly rudiments in vermilion larvae or the reverse. From these embryological approaches, they worked out that the two eye pigments—an orange pigment and a brown pigment—each involved a series of steps leading to the final pigment in each series. They referred to this as "the genetic control of biochemical pathways."[7] They also proposed that a complex molecule is built by a series of enzymatic steps and that a specific gene controls each step. They called this "the one gene–one enzyme hypothesis." This did not, however, tell them the chemical nature of the two pigments or the chemical nature of the intermediate products in each pathway. For that reason, Beadle switched to a fungus, *Neurospora*, and asked the help of a biochemist, Edward L. Tatum (1909–1975), to participate in an analysis of biochemical pathways associated with the synthesis of vitamins and other small molecules in the cell. They induced mutations with X rays and got abundant numbers of strains lacking the ability to make a particular vitamin they were studying. Instead of exotic chromosome balancers and combinations of genes and rearrangements to design stocks, they used survival or nonsurvival of individual colonies first plated on a fully supplied medium containing the vitamin they were studying. If the sampled colony failed to survive on a minimal medium, it was classified as a mutation in the pathway leading up to its synthesis.[8]

Most geneticists welcomed this new approach to genetics using microorganisms. In one sense, it was a fulfillment of Muller's hope in 1921 at the Toronto American Association for the Advancement of Science (AAAS) meetings that geneticists would be able to treat genes as chemicals that could be studied in a test tube. But Muller cautioned that genes were not just committed to a single step in a biochemical pathway, and some genes had multiple effects on different tissues and possible different biochemical processes in the cell.[9] Later such complexities were encountered, but in its basic finding, the one gene–one enzyme theory was a fundamental advance in genetics that ushered in a biochemical age using yeast cells, bacteria, and fungi to relate mutations to metabolism and the synthesis of almost every molecule in the cell.

At Caltech, Norman Horowitz (1915–2005), working with Beadle's group, proposed a tie of the one gene–one enzyme theory to evolution.[10] He proposed a "backwards evolution" for biochemical pathways going from more complex to simpler molecules through a series of enzymatic steps that started first with using the complex molecules present in the earth's early history, then adding a side group or other alteration to a molecule not quite finished. This, in turn, would be used up, and any mutation that could attach another needed component would use what was in the residue until, step by step, an initial small abundant component was the progenitor of the pathway.

Genetic Analysis Reveals Another Type of Position Effect

Clarence P. Oliver (1898–1991) in 1940 discovered an unusual event associated with crossing over while working with the *lozenge* multiple alleles. He found that *lozenge-1* and *lozenge-spectacle,* when mutually heterozygous, occasionally produced an offspring with wild-type round eyes. Unlike the bar eye case, the event was associated with a consistent behavior in which crossovers always carried the same marker placement, suggesting that *lozenge-spectacle* was immediately to the left of *lozenge-1.*[11] Edward B. Lewis (1918–2004) found similar events with the *Star-asteroid* alleles, but was even more rewarded when he studied the *bithorax* series of alleles.[12] These would have thrilled Bateson. Bithorax was a homeotic mutant in which wing, leg, balancer, and abdominal segments could take on cognate roles through mutation. Lewis was able to produce four-winged flies or no-winged flies and eight-legged or four-legged flies by combining different mutations in the segments associated with this *bithorax* region. In Lewis's interpretation, the existence of differentiated duplicated genes had led to a new type of position effect that was revealed by the *cis* or *trans* alignment of these elements.[13] In the *cis* alignment, the double heterozygote was normal in phenotype. In the *trans* alignment, the two normal alleles showed a "pseudoallelism," as Barbara McClintock first called it. They expressed a mutant phenotype. The large number of pseudoallelic genes that had been documented by the mid 1950s suggested to Guido Pontecorvo (1907–1999) that all genes are pseudoallelic, not because they are duplicated and differentiated, but because the crossover event was occurring between two lesions in a common gene.[14] It turned out that both were correct. Sometimes the presence of the mutually heterozygous alleles acted as if they were nonallelic. This function was given a name, "complementation." It was most dramatic in the *dumpy* series of pseudoalleles when an allele with oblique wings was heterozygous with a homologous chromosome bearing the *dumpy* allele expressing thoracic vortices. The *dumpy-oblique/dumpy-vortex* heterozygote was normal in appearance, yet each of these components was allelic in expression when heterozygous with *dumpy* itself, which had both oblique wings and thoracic vortices.[15]

Note how the simple idea of allelism had evolved. In Bateson's era, alleles were variants of the same gene that did not emit crossover-associated reversions to the normal. The discovery by Oliver that this was not universal led to the term "pseudoallelism" for this new phenomenon. But because some of the mutant alleles did not show a mutant expression when mutually heterozygous, the term "complementation" was introduced. Mutant genes could thus form multiple allelic series, pseudoallelic series, and mixtures of expressions that included complementing or mutant expressions in the mutually heterozygous state. Muller called such systems "complex loci." It also led to the term "sublocus" for each of the components mapped in a pseudoallelic locus or system. The term sublocus would horrify mathematicians, who saw a locus on a line as lacking dimensions.[16] At the same time, a fixed sequence of components expressed some sort of position effect because Lewis believed that there were at least two duplicated units involved in each case of pseudoallelism. He attributed the high number of such series to the existence of numerous "repeats," as Bridges called the salivary bands that showed numerous such sets of repeated bands.[17] For Lewis, this was evidence for Muller's hypothesis of gene evolution through primary unequal crossing over as in the bar case. Unlike bar, which was recent, Lewis attributed the proper pairing of the inferred duplicated units to their differentiation over long periods of dipteran evolution. This made the pseudoallelic series physiologically dependent on each other to bring about the position effect. Lewis associated this with an inferred localization of gene products moved along genes corresponding to a pathway that Pontecorvo referred to as "milli-micro-molar reactions."[18] The resolution of how pseudoallelism works came through the development of molecular genetics, which was concurrently developing, in almost explosive growth after the publication of the double-helix model of DNA.

Molecular Biology Shifts Mutation into the World of Nucleic Acids

Nucleic acids reemerged from their 30-year slumber in the mid 1940s, long after Levene's 1913 analysis of the nucleotide composition of DNA. There were suspicions that nucleic acids, especially DNA, were involved in the mutation process because Torbjorn Oskar Caspersson (1910–1997) and Jack Schultz noted that the frequency of induced mutations with ultraviolet radiation exposure went up dramatically at the wavelength associated with the maximum absorption of ultraviolet by DNA.[19] Some argued that this did not necessarily mean that DNA was directly involved. If it served as a scaffolding, the disruption of that framework could lead to a loss of genes or gene activity. Much more convincing were the studies of the Rockefeller Institute group of Oswald Avery (1877–1955), Colin MacLeod (1909–1972), and Maclyn McCarty (1911–2005). They were careful to obtain purified DNA, RNA, protein, and carbohydrates from strains of pneumococcus bacteria (now known as *Diplococcus pneumoniae*) associated with mammalian pneumonia. In their studies with mice, they found that fractions of bacterial cells containing DNA acted as agents directly altering recipient nonvirulent cells into virulent cells, including the formation of the type-specific carbohydrate in the walls of infectious cells that supplied the DNA.[20]

The Rockefeller team called their process "pneumococcal transformation." In 1945, Muller used their report to add a footnote to a lengthy paper on the gene that he was preparing to present to the Royal Society. Muller proposed that fragments of DNA containing one or more genes would enter the recipient bacterium, pair with homologous DNA, and undergo a kind of crossing over to insert the virulence factor gene into the recipient cell.[21] Note how Muller used classical genetics to imagine the process of bacterial transformation as a form of double crossing over. For him, the transition from a line carrying point mutations to a DNA filament composed of nucleotides inserting a small segment carrying a particular gene was no major feat of imagination. Muller's terminology was old, but the transforming event was something new. The gene had acquired a chemical identity in Muller's mind that concurred with the inferences of the Rockefeller group that DNA was the chemical basis for heredity. It was a fulfillment of an expectation he first proposed in 1921 in Toronto. The gene had moved out of the breeder's laboratory and into the test tubes and tools of the biochemist and cell biologist.

Even more remarkable, about the same time in Dublin, the physicist Erwin Schrödinger (1887–1961) was preparing his public Christmas lectures that would appear later the next year in a small but powerfully influential book, *What Is Life?*[22] Ironically, Muller played a significant role in stimulating those lectures, because Muller worked in Berlin in 1932 as a Guggenheim Fellow in the laboratory of Nikolay Timofeef-Ressovsky (1900–1981). Timofeef-Ressovsky added Karl Zimmer (1911–1988) and Max Delbrück (1906–1981) to study the gene from a physicist's perspective. They studied the gene from the perspective of "target theory" to estimate the number and size of genes in a nucleus using X rays to induce the mutations.[23] Muller was using radiation as a tool to investigate gene size, shape, and number. Hitler's rise to the Chancellorship of Germany ended the experiments after a raid on the Institute, and Muller fled to the USSR. Delbrück remained for another two years and continued his explorations of the gene from the outlook of the physicist. His first effort in genetic theory was a failure. He tried to relate quantum theory to the mutation process, using a model not too different from Francis Galton's polygon of instabilities.[24]

Erwin Schrödinger was captivated by this physical approach, and his thesis for the reader of his book was that physics would provide a basis for studying the gene and that this could be accomplished by considering the gene to be a crystal of a special kind. Schrödinger used the term "aperiodic crystal" to describe his image of Muller's belief that a gene could copy its errors and that the copying process was unaffected by the individual point mutations occurring within it. Schrödinger also proposed a way in which genes could be used to store and transmit information. He suggested that they served as a "code script" for biological information used by the cell. The code involved, he argued, need not be elaborate. He compared English language transmission by telegraph using Morse code to his imagined code script. That system used a combination of dots, dashes, and spaces to send the equivalent of 26 letters of the alphabet by the shortness or length of the sound of each dot or dash or interval between sounds.

Muller took Schrödinger to task in a review of his book, not for his physicist's approach to genetics, which he welcomed, but for his invoking a mystical (to Muller) idea of "negative entropy" as the basis for life.[25] For those more materialist-minded scientists who entered the first decade of the molecular revolution, their holy grail was not negative entropy; it was an opportunity to find the structure of the gene and to decode it.[26]

While this was taking place in the postwar years, the hemoglobin molecule took on significant importance. For Max Perutz (1914–2002), it was an opportunity to use X-ray diffraction to study the structure of proteins.[27] The use of X-ray diffraction for complex biological substances began in the 1930s, with fibers like linen or wool providing a challenge. Earlier studies in the 1920s had worked out the structure of long-chain fatty acids. If proteins could be worked out for their amino acid connections and locations, perhaps genes would follow.

At the same time, hemoglobin was taking on significance for another reason. Sickle cell anemia was shown by Anthony Allison (1926–) to be a recessive gene that in the homozygous condition caused a debilitating and often lethal condition in young children who inherited it.[28] It was also associated primarily with people of African descent. Linus Pauling (1901–1994) took an interest in the hemoglobin molecule and, in working out the amino acid content, found that the difference between normal hemoglobin and the mutant hemoglobin involved a tiny difference. Pauling believed that that difference, perhaps as small as a single amino acid difference in a molecule of about 300 amino acids, led to the altered physical properties that clumped the hemoglobin and formed irregular sickle cells. That, in turn, led to their entanglement in capillaries and to the subsequent necrosis of tissue, the stunting of growth, the deprivation of oxygen, and other symptoms of sickle cell anemia. Pauling referred to sickle cell anemia as a "molecular disease."[29] The phrase was brilliant, and Pauling's insight had a dual effect. It made gene mutations and the diseases they caused accessible to biochemists. It also led to an enriched sense of human genetics that would soon be called "medical genetics."

With the development of chromatography, Pauling and Harvey Itano (1920–2010) soon extended the study of hemoglobin. They characterized the molecule as composed of two chains, α and β. It was the β chain that turned out to be carrying what was the result of a point mutation. The digested fragment, studied by "genetic fingerprinting" by Vernon Ingram (1924–2006), indicated that it was a shift from a negatively charged glutamic acid to a neutral valine acid in the number six position of the first fragment of the β chain that was involved in converting hemoglobin A (HbA) into hemoglobin S (HbS).[30] This was consistent with Pauling's expectation that the hemoglobin mutation would turn out to have a positive charge in the β chain. At the same time, Allison, following a suggestion by J.B.S. Haldane (1892–1964), associated the distribution of the HbS allele with the distribution of malaria in the equatorial regions of Africa. Allison proposed an adaptive advantage for the heterozygote, HbA/HbS, in surviving malarial infections; a risk of death from malarial infections for those who were normal, HbA/HbA; and death from sickle cell anemia for those homozygous for it, HbS/HbS. This left adults who were largely heterozygous better opportunities to reproduce, and hence a "balanced polymorphism" was established as the mechanism for the survival of the HbS gene. In North America, however, there was no malarial infection to prop up the HbS allele in the population. Treatment was still palliative, and many children with sickle cell anemia did not live to reproduce. Hence the HbS allele was slowly being removed from North Americans of African descent since their arrival in the slave trade.

The hemoglobin story had a powerful effect on the thinking and teaching of geneticists in the 1950s. They realized that genetics had reached a new stage of maturity in which evolution, point mutation, breeding analysis, biochemistry, anthropology, population genetics, and clinical medicine had become confluent in a single consistent interpretation of this molecular disease.

The Double-Helix Model of DNA Turns Genetics in a Molecular Direction

It was as exciting for geneticists to experience the 1950s as it must have been for geneticists to experience the decade between 1910 and 1920. So much that was new was coming from so many different fields. People, mostly in their 20s or early 30s, who were unknown would do experiments that would make them household names in the community of geneticists. Delbrück had fled Nazi Germany and began the work on bacteriophage genetics that would launch the field of molecular genetics. His students and colleagues explored the relation of DNA and protein (the only living components of bacteriophage). Delbrück's laboratory worked out the life cycle of bacteriophage, demonstrating the infectious stage (a latent stage, as it was originally called) in which the bacteriophage particles were multiplied and assembled, a burst or lysis of the infected cells, and a release of 100 or more mature progeny ready to infect new bacteria.[31] They also discovered mutations in bacteriophage producing smaller or larger plaques or colonies in the lawns of bacteria that appeared rough or turbid. Salvador Luria (1912–1991) learned that he could perform multiple infections of bacteria by two strains and obtain recombinant viruses with both mutant strains present or with the two normal alleles present. The colony counts provided Luria the same opportunity as Morgan's data provided Sturtevant, and a map of bacteriophage point mutations was soon available.[32]

Alfred Hershey (1908–1997) and Martha Chase (1927–2003) used radioactive labeling to demonstrate that DNA enters the bacterium and not the protein of the virus. From this inserted DNA, 100 or more mature viruses emerge after 30 minutes. The viral protein had to be produced by the viral DNA.[33] For most geneticists, this 1952 experiment confirmed the pneumococcal results—that DNA, and not protein, was the genetic material. What was convenient in the bacteriophage studies was the virtually unlimited numbers of viruses they could get to mutate or to recombine. The maps quickly revealed that a gene like *rII* in bacteriophage T4 consisted of a large sequence of points on a line. Seymour Benzer (1921–2007) by 1954 was able to extend that analysis in the rII region. It consisted of two regions, rIIA and rIIB. Each segment had several hundred sites. Most of the mutations were obtained by using chemical mutagens, with ethyl methanesulfonate (EMS) being the most useful agent for this. Benzer referred to this as "genetic fine structure."[34]

In 1953, James Watson and Francis Crick at Max Perutz's Cambridge laboratory succeeded in working out the structure of the DNA molecule, after initial failures.[35] They had competed with Maurice Wilkins (1916–2004) working at the University of London, Rosalind Franklin (1920–1958) independently working at the University of London, and Pauling at Caltech. Watson and Crick used X-ray diffraction of DNA crystals to identify the basic structure (a double helix) and model building (which was Pauling's forte) to work out the pairing associations of purines with pyrimidines.[36] Their resulting molecule had some important genetic features. First, they proposed that their model of the double helix accounted for replication of DNA as a "semiconservative" event—each DNA strand of the double helix replicated by assembling a complementary strand. The mechanism was called "semiconservative replication." Second, the pairing relations accounted for the distributions earlier found by biochemist Erwin Chargaff (1905–2002) and known as Chargaff's laws. Those laws specified that in any given species he studied, there was a ratio of 1:1 for adenines with thymines and for cytosines and guanines. In different species, however, the ratio of A + T/G + C varied. Third, the sequence was aperiodic and thus consistent with Schrödinger's aperiodic crystal concept. The aperiodicity meant that each gene had a unique sequence of nucleotides. This meant that each gene carried a codescript or "genetic code" specifying its product, probably a protein. Fourth, a replacement of a nucleotide might be the basis for point mutations leading to a corresponding replacement of an amino acid in the protein produced by the gene, as Pauling and Ingram demonstrated for the normal and sickle cell hemoglobin results discussed above. This relation of gene mutations to protein alterations assumed what Crick called the "colinearity" of DNA with protein. It was also widely assumed in the mid 1950s that the production of proteins by DNA required an intermediary because genes did not leave the nuclei of cells. But RNA did. Thus RNA was considered the carrier of specificity, information, or genetic coding, into the cytoplasm, where that activity of decoding took place. Because that decoding led to protein formation, this meant that the cell cytoplasm was the place where protein synthesis occurred. In the late 1950s, protein synthesis was identified with the endoplasmic reticulum and its collection of ribosomes.

The story that appears as a narrative in texts did not proceed in a step-by-step logical order as these various findings emerged but were pieced together. A guiding motto was in the mind of geneticists throughout the 1950s. Watson and Crick referred to it by the shorthand phrase DNA → RNA → protein, which was orally described as "DNA makes RNA makes protein." It was described by Crick as "the central dogma" because at the time they had no experimental evidence for it. Pieces of the story came from radioisotope labeling and exposure on photographic emulsions of cells undergoing mitotic replication, of centrifuged fungi undergoing protein synthesis, and of chromatographic distribution of molecules from centrifuged fractions obtained from cells.

More elaborate experiments were used to demonstrate the essentials of molecular genetics. Proof of the semiconservative replication came from the Matthew Meselson (1930–)–Franklin Stahl (1929–) experiments on bacteria fed nutrients with a heavier isotope of nitrogen.[37] The DNA produced could be centrifuged in solution with heavy metal salts, producing density gradient centrifugation separations of timed DNA showing parental and offspring DNA as bands that could be photographed. Their experiments were consistent with the predictions of the double-helix molecule.

The molecular approaches of the 1950s eclipsed the genetics by phenotype and breeding that characterized classical genetics. The basic idea of a mutation as being a small disturbance in a gene or point mutation was not challenged. Benzer's plethora of *rII* mutation sites was assumed to be a map of the gene itself. Genetic fine structure was assumed to correspond to the nucleotides that composed the DNA of the rII region. Molecular geneticists did not discover point mutations, genes, or chromosome maps. Rather, they used those classical genetic concepts and molecularized them. But the phage group suffered from the same limitation as did classical geneticists using fruit flies or maize. They did not know the biochemical functions of the phage mutations. The biochemistry was missing. That limitation was overcome by the work of Milislav Demerec (1895–1966) at Cold Spring Harbor. He used the bacterium *Salmonella typhimurium* and obtained mutations for the sites of some of the genes involved in tryptophan synthesis.[38] Here were cognate genes similar in sequence but slightly different in function, each producing an enzyme associated with a step in the biochemical pathway leading to tryptophan synthesis from its precursors. The biochemical pathway and the participating genes were arranged in sequence. This suggested that genes were tandemly duplicated in some distant past and had differentiated into cognate genes that carried out functions associated with a biochemical pathway. To Demerec and some fruit fly geneticists following his work, this was an explanation for pseudoallelism and its position effect. The pathway was associated with gene sequence. Much later this epigenetic position effect was associated with the messenger RNA (mRNA) fed into ribosomes for producing the enzymes involved.

It may not have been a literal "milli-micro-molar" process, but in bacteria that association of genes with the steps in a pathway allowed products to be synthesized in close proximity. It also was consistent, for fruit flies, to have duplicated genes that had differentiated into cognate functions expressed as a multiple allelic series within which crossing over occurred between duplicated genes as well as (much more rarely) within a particular gene of a complex locus.[39]

References and Notes

1. Bateson W. 1907. Inaugural address: The progress of genetic research. In *Report of the Third 1906 International Conference on Genetics: Hybridization (the Cross-Breeding of Genera or Species), the Cross-Breeding of Varieties, and General Plant Breeding* (ed. Wilks W), pp. 90–97. Royal Horticultural Society, London.
2. Garrod A. 1902. The incidence of alkaptonuria: A study in chemical individuality. *Lancet* **II:** 1616–1620.
3. Olby R. 1974. *The path to the double helix: The discovery of DNA.* University of Washington Press, Seattle.
4. Landsteiner K. 1900. Zur kenntnis der antifermentativen, lytischen und agglutinenierenden wirkungen des Blutserums' und der lymphe. *Zentralbl Bakteriol* **27:** 357–362.
5. Wilson EB. 1896. *The cell in development and inheritance.* Macmillan, New York.
6. Levene P, London ES. 1929. On the structure of thymonucleic acid. *J Biol Chem* **83:** 793–802. Although Levene proposed a tetranucleotide model for DNA about 1909–1912 (known then as thymonucleic acid), he did not provide a diagram for it. He thought of DNA as a set of the four bases A, T, G, and C in some sort of repetitive nucleotide sequence. He did not consider nucleic acids to be significant for playing a hereditary role.
7. Beadle GW, Ephrussi B. 1935. Transplantation in *Drosophila. Proc Natl Acad Sci* **21:** 642–646.
8. Beadle GW, Tatum EL. 1941. Genetic control of biochemical reactions in *Neurospora. Proc Natl Acad Sci* **27:** 499–506.
9. Muller's caution stemmed from his work on Beaded and Truncate. He believed that there were so many activities of individual genes in metabolism and development that these were unlikely to reflect an exclusive model of one gene and one enzyme as the basis of observed phenotypes in fruit flies. To Muller, the Beadle and Tatum findings were a component of a more complex system of gene interactions.
10. Horowitz N. 1945. On the evolution of biochemical syntheses. *Proc Natl Acad Sci* **31:** 153–157.
11. Oliver CP. 1940. A reversion to wild-type associated with crossing over in *Drosophila melanogaster. Proc Natl Acad Sci* **26:** 452–454.

12. Lewis EB. 1951. Pseudoallelism and gene evolution. *Cold Spring Harb Symp Quant Biol* **16:** 159–174.
13. Lewis E. 1955. Some aspects of position pseudoallelism *Am Nat* **89:** 73–89.
14. Pontecorvo G. 1952. Genetic formulation of gene structure and gene action. *Adv Enzymol Relat Subj Biochem* **13:** 121–149. Pontecorvo G. 1955. Gene structure and action in relation to heterosis. *Proc R Soc Lond B Biol Sci* **144:** 171–177.
15. Carlson EA. 1959. Allelism, pseudoallelism, and complementation at the *dumpy* locus of *D. melanogaster. Genetics* **44:** 347–373.
16. It cost me a job at the University of Missouri in 1958 when I was a candidate. One of the members of the committee examining me grilled me on why I used "sublocus" as a term in genetics when a point, to a mathematician like him, had no dimension and thus could not be divided. I felt like the idiot he thought me to be.
17. Bridges CB. 1936. Correspondence between linkage maps and salivary chromosome structure as illustrated in the tip of chromosome 2 R of *Drosophila melanogaster. Cytologia* **Fuji Jubilee volume:** 745–755.
18. Pontecorvo G. 1952. Genetic formulation of gene structure and action. *Adv Enzymol Relat Subj Biochem* **13:** 121–149. Also, for a more extended discussion, see Chapter 21, "Pseudoallelism versus intragenic recombination," pages 184–195 in Carlson EA. 1966. *The gene: A critical history.* Saunders, Philadelphia.
19. Caspersson T, Schultz J. 1950. Cytochemical measurement in the study of the gene. In *Genetics in the 20th century* (ed. Dunn LC), pp. 155–172. Macmillan, New York.
20. Avery OT, Macleod CM, McCarty M. 1944. Studies on the chemical nature of the substance inducing transformation of Pneumococcal types: Induction of transformation by a desoxyribonucleic acid fraction isolated from pneumococcus type III. *J Exp Med* **79:** 137–158.
21. Muller HJ. 1945. The Gene Pilgrim Trust Lecture. *Proc R Soc Lond B Biol Sci* **134:** 1–37.
22. Schrödinger E. 1945. *What Is life?* Cambridge University Press, Cambridge.
23. Timofeef-Ressovsky NV, Zimmer EG, Delbrück M. 1935. Über die natur der Genmutation und Genstrucktur. In *Nachrichten aus der Biologie des Wissenschaften Göttingen,* Vol. 1, pp. 234–241.
24. Conversation with Max Delbrück at UCLA. Delbrück was not impressed with his own model of the quantum gene and called it "a silly piece of work."
25. Muller HJ. 1936. The need of physics in the attack on the fundamental questions of genetics. *Sci Mon* **44:** 210–214.
26. Cairns J, Stent G, Watson JD. 1966. *Phage and the origins of molecular biology.* Cold Spring Harbor Laboratory, Cold Spring Harbor, NY.
27. Ferry G. 2007. *Max Perutz and the secret of life.* Cold Spring Harbor Laboratory Press, Cold Spring Harbor, NY.
28. Allison AC. 1954. Notes on sickle-cell polymorphism. *Ann Hum Genet* **18:** 39–51.

29. Pauling LH, Itano HA, Singer SJ, Wells IC. 1949. Sickle cell anemia, a molecular disease. *Science* **110:** 543–548.

30. Ingram V. 1957. Gene mutation in human hemoglobin: The chemical difference between normal and sickle cell hemoglobin. *Nature* **180:** 326–328.

31. Ellis EL, Delbrück M. 1939. The growth of bacteriophage. *J Gen Physiol* **22:** 365–384.

32. Luria S, Delbrück M. 1943. Mutations of bacteria from virus sensitivity to virus resistance. *Genetics* **28:** 491–511.

33. Hershey AD, Chase M. 1952. Independent functions of viral protein and nucleic acid in growth of bacteriophage. *J Gen Physiol* **36:** 39–56.

34. Benzer S. 1961. On the topography of the genetic fine structure. *Proc Natl Acad Sci* **47:** 403–415.

35. Watson JD, Crick FHC. 1953. Molecular structure of nucleic acids. *Nature* **171:** 737–738.

36. Many accounts are available from the participants or their close associates. See Watson JD. 1968. *The double helix.* Athenaeum, New York. Sayre A. 1975. *Rosalind Franklin and DNA.* Norton, New York. Wilkins M. 2003. *The third man of the double helix: The autobiography of Maurice Wilkins.* Oxford University Press, New York. Crick FHC. 1988. *What mad pursuit: A personal view of discovery.* Basic Books, New York. Maddox B. 2002. *Rosalind Franklin: The dark lady of DNA.* HarperCollins, London.

37. Meselson M, Stahl FW. 1958. The replication of DNA in *Escherichia coli. Proc Natl Acad Sci* **44:** 671–682.

38. Demerec M. 1957. Structure and arrangement of gene loci. Proceedings of the International Genetics Symposia, Tokyo and Kyoto. *Cytologia (Tokyo)* (Supplement) **1957:** 20–31.

39. Op. cit. (note 15).

7

Mutation in Relation to Gene Structure

THE INTRODUCTION OF MENDELISM HAD generated a set of new terms to describe mutations, and so too did the chromosome theory of heredity. Now the introduction of biochemical analysis of DNA and protein also introduced a flood of new terms. Some of the terms were molecular replacements of classical genetic terms. One could speak of a nucleotide base-pair change or "base substitution" in the sequence of DNA instead of a point mutation in a specific gene. The newer term was usually more informative. Not all the new terminology persisted. Seymour Benzer, whose initial background was physics, used the physicist's penchant for adding the suffix " -on" to fundamental units in the atom (e.g., neutron, positron, electron). By extending the idea to the gene, he came up with the "cistron" as the unit of function, the "recon" as the unit of recombination, and the "muton" as the unit of recombination.[1] Philosopher–physicist Percy Bridgman (1882–1961) had applied a utility test to definitions that appealed to Benzer. They were an expression of what Bridgman called "operational definitions."[2] Bridgman believed that a concept was best described by how it was used and not by some innate quality. His philosophy was called operationalism. Virtually no geneticist used the terms muton and recon extensively. But the term cistron caught on for a while, and many geneticists using fruit flies were referring to their pseudoalleles as cistrons or the entire collection of pseudoalleles as consisting of a "polycistronic system." By the 1970s, geneticists lost interest in using operational definitions and shifted back to definitions that conveyed mechanism or composition for fundamental units of heredity.

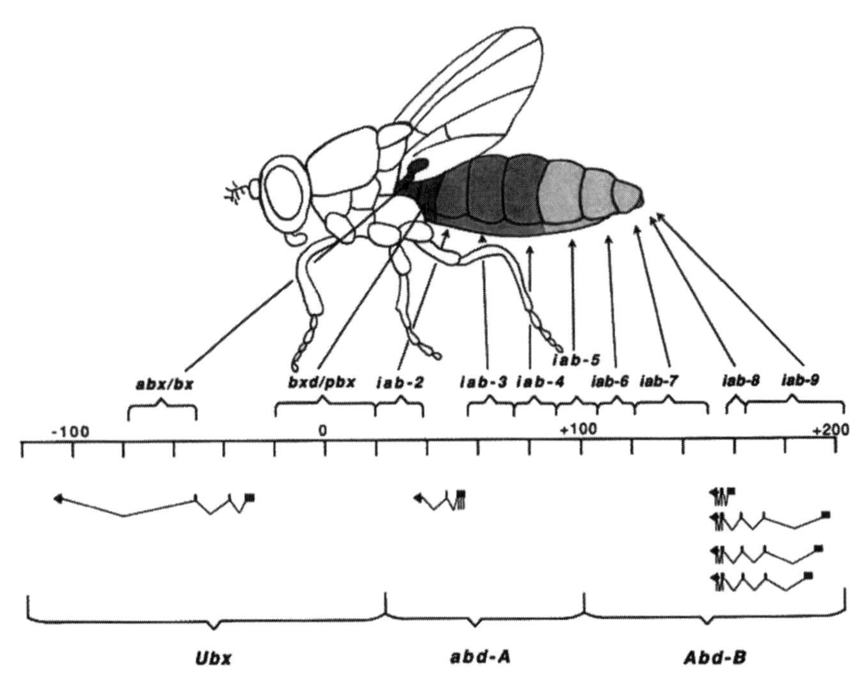

The origins of evo-devo through the analysis of bithorax. Edward B. Lewis used pseudoallelism at the bithorax locus to map mutant alleles of the bithorax region. These alleles had either a deficit or an excess of the anterior or posterior concentration of a morphogen associated with wing, haltere, or abdominal segment. Lewis succeeded in bringing classical genetic techniques to a developmental genetic interest in how organ systems differentiated and what controlled their formation. (Reprinted, with permission, from Duncan I, Montgomery G. E.B. Lewis and the bithorax complex: Part II. From cis-trans *test to the genetic control of development.* Genetics **161:** *1–10. © Genetics Society of America.)*

Genetic Fine Structure Relates Gene Structure to Mutagenesis and the Genetic Code

Although Benzer's terminology may have been inherently flawed, that was not the case for his finding of genetic fine structure and the mapping of the *rII* locus. Geneticists immediately recognized that Benzer, and virtually simultaneously, Milislav Demerec, had related the gene to the DNA molecule by this detailed mapping of point mutations.[3] So powerful were Benzer's findings on the field of genetics that most were puzzled that his work never merited a Nobel Prize. Benzer and his students also discovered in their mapping process that some sites showed many recurrences of mutation that went beyond Poisson fluctuations. They called such sites "hot spots." Benzer's work also set the stage for comparative mutagenesis using a variety of chemical agents to see if some sites were preferentially affected. Much of this work was done by Benzer's laboratory with Ernst Freese (1925–1990) and by Francis Crick's laboratory with Sydney Brenner (1927–). The Benzer–Freese approach was focused on the comparative mutagenesis of genes.[4] The Crick–Brenner approach was focused on using molecular mutations as a tool to understand the genetic code.[5]

There were several changes possible with chemical mutagens like ethyl methanesulfonate, formaldehyde, or nitrous acid. These acted by altering a single nucleotide in a base pair of a nucleotide sequence. The alteration of one pyrimidine to a different pyrimidine could shift a thymine site to a cytosine site after replication. Such changes were called "transitions." If the cytosine site was changed to guanine or adenine, the mutational change (from pyrimidine to purine) was called a "transversion." Very different were chemical agents with multiple ring structures such as quinacrine, proflavine, or atabrine that were found useful for preventing malarial infections. Such agents were called "intercalating agents" because they were thought to jam between base pairs, alter the rotation of the DNA molecule, and lead to minute deletions or duplications of bases, during the replication process. Such mutations were known as "frameshift mutations" because they more profoundly disturbed gene function as a consequence of the altered reading of the genetic code.[6] That difference was made clear when the genetic code was recognized as involving a sequence of three base pairs to specify a single amino acid. The first person to suggest this did so on theoretical grounds. Physicist George Gamow (1904–1968) argued that at least three base pairs were needed to specify 20 amino acids.[7] One base pair had four possible configurations: AT, TA, GC, or CG. A sequence of two base pairs had 16 possible configurations. For one strand, that could be AA, AT, AG, AC, TT, TA, TG, TC, GG, GA, GC, GT, CC, CA, CG, or CT. That was still four short of 20. But if a sequence of three nucleotide pairs was necessary to specify one amino acid, there were $64 - 20 = 44$ combinations that were in excess. This led to lots of speculation about genetic codes with 44 other roles or subject to silence in the coding process. The sequence of three nucleotide pairs was called a "triplet" or a "codon."[8]

The effect of a frameshift mutation was the conversion of all amino acids following any departure from an initial triplet state into a "gibberish" that destroyed the function of the gene and made it lose its antigenic properties. Crick and Brenner primarily did the analysis of intercalating agents. They used the phrase "modulo-3" to designate the new mutant state. If it was a loss or gain of 3, it acted like a classic point mutation or base substitution or minute rearrangement that otherwise did not alter the gene product. If it was a "modulo-2" or "modulo-1" event, the subsequent reading frame would alter almost all subsequent amino acids in the protein, or it could abruptly terminate if one of the triplets turned out to be an unexpected stop signal downstream from the event. Transitions or transversions, worked out primarily by Benzer and Freese, in contrast, could alter a gene, but its protein product retained reactivity with antibodies. They were called "CRM mutations" for having "cross-reacting material" as their genic product. Although they tried to use molecular genetics to break the genetic code, Brenner and Crick did not succeed. They were scooped because the genetic code was solved by chemical means.

The Genetic Code Is Solved by Direct Chemical Means

Severo Ochoa (1905–1993) devised methods to synthesize RNA.[9] It was widely believed from about 1950 on that RNA was the means by which information in the genes reached the cytoplasm. This type of RNA was called messenger RNA (mRNA). Marshall Nirenberg (1927–2010) and Heinrich Matthaei (1929–) created synthetic mRNA using Ochoa's methods.[10] They introduced this mRNA into a cell-free system they devised for protein synthesis. Thus poly-uracil, the sequence UUUUUUUUUU. . . , fed into a test tube with components for protein synthesis (ribosomes, amino acids, and other cell-free factors) would produce proteins that contained one repeated amino acid, phenylalanine (Phe), and thus a peptide, phe.phe.-phe.phe.phe.phe. . . . By using two nucleotides with one more common than the other, such as four Us to one C, they would get mostly UUU, then two Us and a C (UUC UCU CUU), and then two Cs and a U (CCU, CUC, CCU), and rarest of all, chance sequences of three in the synthetic mRNA, would be those that were CCC. These approaches were time-consuming but productive, and thus all 64 codons were worked out by Nirenberg's group, Ochoa's group, and H. Gobind Khorana's (1922–) group.[11] What they found was unexpected: Most amino acids can be coded by more than one triplet sequence. Four of the 64 combinations serve as signals for either starting the sequence (AUG) or for ending it (UAA, UAG, and UGA) in a gene. These non–amino acid codons were called "stop signals" if they terminated the reading of the mRNA in a ribosome. The process of copying DNA into mRNA was called "transcription." The process of decoding proteins from nucleotide triplets in mRNA was called "translation."

Crick, on theoretical grounds, proposed another category of RNA to supplement the RNA found in ribosomes and the RNA produced by the genes known as mRNA. He argued for "adaptor molecules" that would carry the amino acids to the ribosome so they could be placed into the growing sequence of amino acids in a protein. These acquired the name "transfer RNA" (tRNA).[12]

As often happens with the discovery and introduction of a new idea or a new process, complications arise that could not be anticipated at the time of its first introduction. One of these was the finding by Phillip Sharp (1944–) and independently by Richard Roberts (1943–) that genes in eukaryotic cells differed in structure from genes found in bacteria. In bacteria, the genes containing a given number of nucleotides produced proteins with one-third the number of amino acids. Thus a 300-base-pair gene would produce a 100-amino-acid protein. This was known as "collinearity." Crick and other geneticists considered this as universal for all living things. To their surprise, the findings of Sharp and Roberts revealed that eukaryotic genes (or some viruses associated with animal infection like SV40) were usually about 10 times the size of bacterial genes, but their protein products were considerably smaller.[13] When they compared the sequence of nucleotides in eukaryotic genes to the protein product, they noted that the coding regions specifying the amino acids were discontinuous, broken into chunks called "exons." Between adjacent exons were sequences of intervening DNA called "introns." This created problems for those who were studying protein synthesis. How were these exons read? This led to the discovery of processing enzymes, including ribozymes, which are nonprotein enzymes, that snipped out the introns from an initial sequence of pre-messenger RNA including both introns and exons.[14] The excision process required additional enzymes to remove the introns and splice together and modify the set of exons to form the mRNA that would be used for translation into a protein. The process of modifying the transcribed RNA from DNA introduced another term, "splicing," to the vocabulary of geneticists working with gene structure and function.

The addition of introns to genes and the addition of excision and splicing enzymes in the process of transcription had several major implications that were studied. First, the act of splicing could lead to errors and generate mutations. Second, the introns not producing the protein associated with the gene were largely neutral with respect to natural selection; and mutational changes in them could be used to trace such things as mutation rates in the history of a species or of higher taxa. Third, the exons could serve as "domains" that could be duplicated or shuffled, and this greatly extended the number of genes beyond the duplication of an entire gene or a more substantial genetic region as in the bar duplication.[15] The implications of domains instead of proteins clarified some of the objections Muller and others had to the universality of the one gene:one enzyme model of George Beadle and Edward L. Tatum. The diverse effects on organs or time of expression of traits in a pleiotropic gene could be interpreted satisfactorily as a one gene:one domain model. The domain shuffling then becomes an epigenetic effect and greatly increases the number of protein products from a smaller number of genes producing these domains. This also solves the problem of having far fewer genes (about 25,000) than originally anticipated for the human genome. What it does not solve is the shuffling process and how epigenetic rules determine which domains are combined. This epigenetic model also fails to show how the process is regulated. Those aspects will no doubt occupy much of genetics in the 21st century.

Tools Are Introduced for Sequencing Macromolecules

These major events in the history of molecular genetics depended on new tools. Chief among these were the tools for sequencing proteins, mostly worked out in the 1940s and early 1950s.[16] Simple molecules like peptide growth hormones were among the first to be sequenced. More complex ones like hemoglobin required much more effort, including the X-ray diffraction analysis that yielded the sequence of that molecule.[17] Chromatography of digested proteins into a small number of fragments, "peptide fingerprints," was used for a direct biochemical approach to sequencing.[18] The final tool was sequencing DNA. Two approaches were used. One was by

Fred Sanger's (1918–) group in Cambridge, England,[19] and the other was by Walter Gilbert's (1932–) group[20] at Harvard. These relied on gels that separated fragments of different sizes and that were tagged with a marker. The gel chromatography separation could be photographed and the base pairs read off in sequence.

Reading genes had immense implications for medical genetics and the study of human gene mutations. It had immense implications for the field of phylogenetic evolution, where DNA sequences of cognate species and more distant ancestors could be compared. The protein approach for the hemoglobin molecule was richly rewarding for tracing the migration of the sickle cell mutation and distinguishing it from dozens of other mutant occurrences throughout the world. It also revealed how normal human hemoglobin differed from those of other primates, other mammals, and all animal life that used hemoglobin for carrying oxygen to tissues. When DNA could be read, the results were even more revealing about the history of migration of populations carrying specific mutations.

Classical genetics was primarily an academic discipline. There were, nevertheless, many extensions from the universities to field stations and commercial applications of classical genetics in the food industry and horticulture. But molecular genetics provided enormous opportunities to manipulate DNA and shift genetics into a field similar to that of engineering. It was no surprise that when "restriction enzymes" were first found by Salvador Luria and seen as a defense to snip foreign DNA into harmless fragments, Herbert Boyer (1936–), Paul Berg (1926–), and others immediately saw the potential of these restriction enzymes as genetic tools that could conveniently add or shift genes from one species into another.[21] The tools for synthesizing genes to order and plugging in amino acid changes that might make a product more effective for commercial or health benefits also had a dramatic effect on the careers of those getting PhDs in genetics. Few graduate students in the 1940s or early 1950s considered options other than academic ones for their PhD in genetics. They hardly existed in industry. A few jobs were available in agriculture or animal breeding, but virtually none were in the pharmaceutical industry or in hospitals.

But in the 1970s, as these tools for sequencing genes and manipulating genes began to enter the scientific journals, many of the scientists involved entered a partnership with existing corporations desirous of finding new drugs, human biological products (e.g., insulin, human growth hormone, clot-dissolving enzymes), and modified crops, among hundreds of industrial, medical, and agricultural values from oil spill degradation to synthetic fuels.

Hundreds of new companies using DNA technology mushroomed in the 1980s in a race to produce new products.

Whereas most genes were thought to encode the sequences for enzymes, many other proteins were identified as products of genes. Some were structural components of cellular organelles. All of the organelles had membranes with their own specific proteins. These were harder to isolate and identify because of their essential nature to the cell. Without them, cells would not divide. By necessity almost all of them would be products of active genes and expressed from the time of fertilization. François Jacob and Jacques Monod identified a third class of proteins in the late 1950s.[22] They called these "regulatory genes." Their existence was revealed from the study of bacterial metabolism.

The Operon Model of Jacob and Monod Introduces Gene Regulation

For some time, it was known that microbial cells would shift production from one enzyme capable of digesting a sugar like glucose to another sugar like lactose if the medium on which the bacteria fed were deprived of glucose but supplied with lactose. This was known as "adaptive enzyme formation."[23] There was a debate on whether this involved mutations to lactose digestion selected by survival on the appropriate medium, or if the cells en masse shifted their metabolism after exposure to the sugar. The latter turned out to be true, and Jacob and Monod worked out how genes could be turned on or off during metabolism. They demonstrated that one or more genes could be involved by regulation through an "operon" serving as a genetic on or off switch. It included the gene for making the enzyme (or a set of tandem genes for several related enzymes) controlled by one operon that was responsive to a regulatory product. That regulator was often a protein associated with a metabolic product such as a simple sugar. The interaction of the regulatory protein and the operon resulted in turning on the transcription of the genes controlled by the operon.

It was not just metabolism that could be regulated. Jacob and Monod recognized that the life cycle involved a series of events in which genes were turned on or turned off to bring about embryonic development, including tissue and organ formation as well as body plans such as head–tail orientation, dorsal–ventral orientation, and the development of features such as limbs, genitalia, and sensory organs. While Jacob and Monod worked on bacteria, many other geneticists began a search for such regulatory genes in the development of fruit flies and other multicellular animals and plants. Ed Lewis, using the *bithorax* series of pseudoalleles, was quick to recognize that his system might be useful for this analysis. He studied insect physiology and insect embryology to relate his *bithorax* mutations to the life cycle of the fruit fly. He envisioned insect development as consisting of the formation of a series of units or boxes wherein the anterior and posterior segments had different gradations of products. If he combined gene mutations that contained a surplus or deficit of these products, he would obtain the formation of abdominal segments, balancers, wings, or legs appropriate to the content in the boxes assembled by genetic recombination.[24]

Lewis did not know the molecular nature of the class of genes involved in this process. That was worked out by the discovery of a class of genes that lead to duplication of parts, called *hox* genes. They were named by Walter Gehring (1939–) in 1983, a Swiss embryologist, and independently by Matthew Scott (1953–), an American.[25] The term "hox" is an acronym for the term "homeobox," which describes a nucleotide sequence found in these genes, and for the observation that such *hox* genes occur in clusters. The *hox* genes are associated with body segmentation, and each segment controlled by a *hox* gene has anterior–posterior (head or tail) differentiation through the gradients of developmental components in them. *hox* genes are found in multicellular animals throughout the animal kingdom. The discovery of *hox* genes shifted the story that Lewis worked out from genetic analysis to a molecular story of *hox* genes and other regulators that are involved in patterning, that is, the body plan, symmetry, and similar activities that can be described as recognition of position. This was true for the *bithorax* pseudoallelic series and mutations like *antennapedia* (in which the antennae on the head of the fly are converted into legs) expressing what Bateson called "homeosis."[26] These discoveries also launched the merging of embryology and the study of evolution, the original dream of Darwin to find in developmental biology a clue to the evolution of higher taxonomic categories. That merger is today a new field of biology called "evo-devo."[27]

References and Notes

1. Benzer S. 1957. The elementary units of heredity. In *The chemical basis of heredity* (ed. McElroy WD, Glass B), pp. 70–93. Johns Hopkins University Press, Baltimore.
2. Bridgman P. 1927. *The logic of modern physics.* Macmillan, New York.

3. Demerec M. 1957. Structure and arrangement of gene loci. Proceedings of the International Genetics Symposia, Tokyo and Kyoto. *Cytologia (Tokyo)* (Supplement) **1957:** 20–31.

4. Benzer S, Freese E. 1958. Induction of specific mutations with 5-bromouracil. *Proc Natl Acad Sci* **44:** 112–119.

5. Brenner S, Barnett L, Crick FHC, Orgel A. 1961. The theory of mutagenesis. *J Mol Biol* **3:** 121–124.

6. In fruit flies, intercalating agents like acridine mustard produced *lv, ol,* and *olv* alleles almost exclusively. But EMS and other nonintercalating mutagens produced a spectrum of alleles—*lv, ov, olv, ov, o,* and *v*. See Carlson EA, Sederoff R, Cogan M. 1967. Evidence favoring a frame shift mechanism for ICR-170 induced mutations in *Drosophila melanogaster. Genetics* **55:** 295–313.

7. Gamow G. 1954. Possible relation between deoxyribonucleic acid and protein structure. *Nature* **173:** 318. doi: 10.1038/173318a0.

8. Crick FHC. 1958. On protein synthesis. *Symp Soc Exp Biol* **12:** 138–163.

9. Grunberg-Manago M, Ochoa S. 1955. Enzymatic synthesis and breakdown of polynucleotides; polynucleotide phosphorylase. *J Am Chem Soc* **77:** 3165–3166.

10. Nirenberg MW, Matthaei JH. 1961. The dependence of cell-free protein synthesis in *E. coli* upon naturally occurring or synthetic polyribonucleotides. *Proc Natl Acad Sci* **47:** 1588–1594.

11. Khorana HG. 1960. Synthesis of nucleotides, nucleotide coenzymes and polynucleotides. *Fed Proc* **19:** 931–941.

12. Op. cit. (note 8).

13. Bork J, Sharp PA. 1978. Spliced early mRNAs of simian virus 40. *Proc Natl Acad Sci* **75:** 1274–1278.

14. Roberts RJ. 1978. Intervening sequences excised in vitro. *Nature* **274:** 530. doi: 10.1038/274530a0.

15. Gilbert W. 1978. Why genes in pieces? *Nature* **271:** 501. doi: 10.1038/271501a0.

16. du Vigneaud V, Ressler C, Swan JM, Roberts CW, Katsyannis PG. 1954. Oxytocin synthesis. *J Am Chem Soc* **76:** 3115–3118.

17. Perutz M. 1960. Structure of hemoglobin. *Brookhaven Symp Biol* **13:** 165–183. Sanger F. 1952. The arrangement of amino acids in proteins. *Adv Protein Chem* **7:** 1–66.

18. Ingram VM. 1957. Gene mutation in human hemoglobin: The chemical difference between normal and sickle cell hemoglobin *Nature* **180:** 326–328.

19. Sanger F, Nicklen S, Coulson AR. 1977. DNA sequencing with chain-terminating inhibitors. *Proc Natl Acad Sci* **74:** 5463–5467.

20. Maxam A, Gilbert W. 1977. A new method for sequencing DNA. *Proc Natl Acad Sci* **74:** 560–564.

21. Luria SE. 1979. The recognition of DNA in bacteria. *Sci Am* **222:** 88–92. Cohen SN, Chang ACY, Boyer HY, Helling RB. 1973. Construction of biologically functional bacterial plasmids in vitro. *Proc Natl Acad Sci* **70:** 3240–3244.
22. Jacob F, Monod J. 1961. Genetic regulatory mechanisms in the synthesis of proteins. *J Mol Biol* **3:** 318–356.
23. Monod J, Cohen-Bazire G. 1953. L'effet d'inhibition spécifique dans la biosynthêse de la tryptophane-desmase chez *Aerobacter aerogenes. C R Acad Sci* **236:** 530–532.
24. Lewis EB. 1978. A gene complex controlling segmentation in *Drosophila. Nature* **276:** 565–570.
25. Gehring WJ. 1987. Homeoboxes in the study of development. *Science* **236:** 1245–1252. Scott GA, Goldsmith LA. 1993. Homeobox genes and skin development: A review. *J Invest Dermatol* **101:** 3–8.
26. Hughes CL, Kaufman TC. 2002. Hox genes and the evolution of the arthropod body plan. *Evol Dev* **4:** 459–499.
27. Carroll S. 2003. *Forms most beautiful: The new science of evo-devo and the making of the animal kingdom.* Norton, New York.

8

Mutation in Relation to Evolution

AS THE 20TH CENTURY BEGAN, EVOLUTION was believed to involve fluctuating variations that were selected gradually. At least, that was the consensus that orthodox Darwinians shared following Darwin's model of 1859. The biometrics group headed by W.F.R. Weldon and Karl Pearson rendered their version of Darwin's model of evolution by presenting sets of mathematical curves representing the variations of a trait. William Bateson opposed this effort as mere description without insight, and he hoped to find discontinuous changes in heredity because he found an abundance of evidence for them in the published records of variations among living things. Today we identify Bateson's variations as largely those of meristic and homeotic mutations, the meristic consisting of duplicated parts like extra digits in hands or feet and the homeotic as displaced developmental expression like Ed Lewis's *bithorax* complex. Hugo de Vries also sought discontinuous traits, and he imagined that these would follow the model of his *Oenothera* mutations that actually generated new species. The work that Bateson did after the rediscovery of Mendelism focused not on homeotic mutations but on the standard point mutations and changes that breeders had selected for generations with an eye on commercial utility or aesthetic appeal. Those did not lead him to evolutionary insights. Similarly, de Vries hoped his finding in *Oenothera* would appear in other plants or animals. It did not. His primroses were anomalies, a side issue in evolution to be worked out by cytogenetic techniques in the 1920s and 1930s.[1] Finally, there were Lamarckian biologists who could not let go of the directing effect of the environment in generating beneficial mutations rather than selecting the mutations significant to evolution. They looked on August Weismann's objections to Lamarckism as superficial because mutilations were not fundamental environmental influences on heredity. Temperature, food, climate, and other

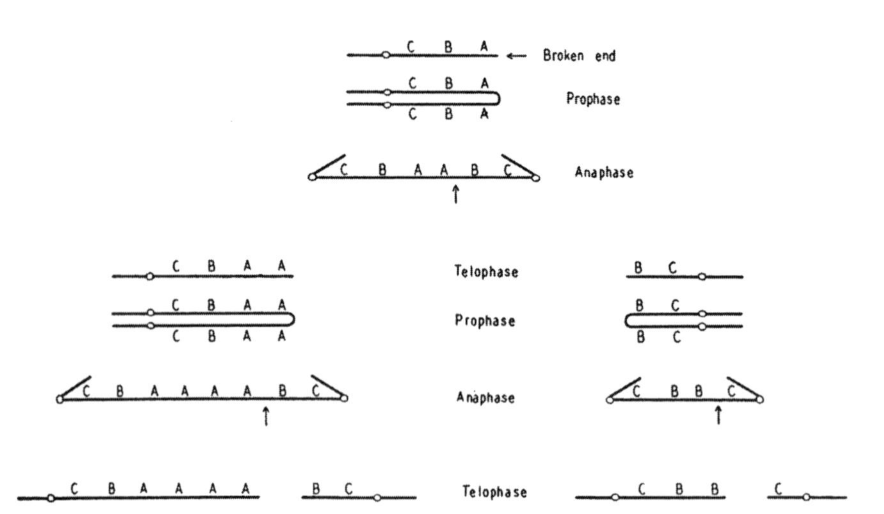

The breakage–fusion–bridge cycle. Barbara McClintock first reported the induction by X rays of the breakage–fusion–bridge cycle in maize while she was in Missouri. She also noted its spontaneous appearance in that species. Here she shows how it works with a diagram. The same finding was reported independently by Guido Pontecorvo and H.J. Muller in Edinburgh using fruit flies. (Reprinted, with permission, from McClintock B. 1941. The stability of broken ends of chromosomes in Zea mays. Genetics **26:** *234–282. © Genetics Society of America.)*

realities of day-to-day life were important, not deliberate or accidental cuttings by knives and other mechanical objects.

Morgan's Group Accepts and Interprets Darwinian Gradualism in Evolution

Only one approach succeeded in supplementing or replacing the view that Darwinian fluctuations mattered. This was the approach of Thomas Hunt Morgan's group, wedded to the chromosome theory of heredity using the fruit fly *Drosophila melanogaster*. It was a better model because the units of heredity, called genes after 1909, could be associated with the chromosomes. They could be mapped and assumed to be aligned linearly from tip to tip of each chromosome. Morgan and his students provided a mechanism for their Mendelian and modified X-linked transmission. They also amassed evidence of when mutations occurred in the life cycle. They demonstrated the existence of multiple allelic series. They characterized mutations for a variety of effects including bicolorism (later to be called dosage compensation), autonomy or nonautonomy in their tissues, reverse mutations (especially the eosin case), pleiotropism, and noncontamination in the heterozygous state over several generations. They also identified chief genes and modifiers to account for the variability of traits, the very variations that Darwin would have called fluctuating. By the early 1920s, they had evidence of a spontaneous mutation frequency for a category of mutations they described as X-linked recessive lethals. Their findings, they claimed, were consistent with Darwinian natural selection, and they forced the confluence between genetics and evolution.

There were two features that were not initiated by the Columbia fly lab. They did not apply mathematics to their theoretical insights. They used statistics, of course, to verify consistency with Mendelian expectations. No student taking a genetics course between 1920 and 1960 would have escaped examination questions involving chi square analysis of data to see if traits were Mendelizing or not. But they offered no models of how genes acted in a population in which selection was taking place. Nor did they stress what later became population genetics. When these new approaches opened up in the first decades of the 20th century, they welcomed that approach because it was consistent with the Darwinian theory they adopted. They argued that point mutations arose randomly. Most point mutations were harmful whether in homozygous or heterozygous condition. Those were eliminated by natural selection. In a few cases, beneficial changes arose, by the same random mechanism producing other point mutations. These were selected and survived and reproduced over numerous generations, infiltrating the population. It was mathematics that determined how survival worked in Darwinian evolution, and it was mathematics that showed how genes were distributed in populations, small and large, with different evolutionary outcomes.

Two major studies led the way. The first was the independent discovery of what is today called the Hardy–Weinberg equation, sometimes represented genetically as AA + 2Ab + bb = 1, but formally as the algebraic equation $p^2 + 2pq + q^2 = 1$, where p is the frequency of the A allele and q is the frequency of the a allele at a given locus. The expansion arises from the assumed random mating of individuals in a population whose gametic contributions are a + b = 1, where a is the dominant allele, and b is the recessive allele. When Reginald C. Punnett (1875–1967) raised his concern about a dominant mutation swamping out normal recessive members of a population, he asked his colleague Godfrey Hardy (1877–1947) at Cambridge University if he could provide a mathematical solution to that problem. In 1908, Hardy showed, with very simple mathematics, that the frequency of a gene in a population would remain constant if the population was large and if there were no selective advantage or disadvantage to the mutation studied. Independently, that same law was also worked out by Wilhelm Weinberg (1862–1937) that same year.[2] In 1921, Hermann Joseph Muller attended a paper given by Charles Danforth (1883–1969) of Johns Hopkins University. Danforth presented a series of equations that could be used to follow the fate of a gene in a population.[3] He assigned different degrees of "impairment" associated with a putative new mutation and showed how long it would take (its "persistence") for that gene under that constraint to be eliminated from the population. Eventually Danforth's equations would be supplemented by Muller to generate his concept of "genetic load," in which a population tolerates a certain amount of spontaneously arising mutations, eliminating as large a percentage through "genetic deaths" as would be arising randomly at a fixed rate.[4] The load represents that balance between origination of new mutations and elimination of an equivalent amount of "detrimental" mutations. That balance can shift up or down depending on the selective forces acting on a population.

Both of these approaches were not based on sophisticated mathematics. That sophistication came from three sources mostly between 1919 and 1930 through the contributions of Ronald A. Fisher (1890–1962) and

J.B.S. Haldane in Great Britain and Sewall Wright (1889–1988) in the United States. They are founders of the population genetics branch of heredity, and their equations were significant for investigating models of evolutionary success or failure for genes or combinations of genes in populations and how the size of populations alters outcomes because of phenomena like "founder effects" associated with small colonizers and "genetic drift" associated with the breaking up of a larger population by environmental changes (such as island formation from an isthmus or other natural barriers).[5] Although most geneticists accepted population genetics as a valued part of the field of genetics, only a specialized few could read these highly mathematical papers. Some naturalists, including the evolutionary biologist Ernst Mayr (1904–2005), rejected this approach as "beanbag genetics" and felt that only fieldwork could determine how variations played out in real environments. Haldane was an enthusiast for this mathematical approach, but Mayr was unimpressed; nonetheless, both had admiration for each other's work and personalities, and they based their dispute on reasoned arguments in their correspondence.[6]

Mayr stressed species formation as arising from isolation of two populations leading to a permanent biological sterility if members of the two groups came into sexual contact. His work was primarily done through field studies of birds. Mayr played a major role in the "new synthesis," as Julian Huxley (1887–1975) called it, for the merger of classical genetics, population genetics, systematics, formal biology, and paleontology that shaped mid-20th century studies on evolution.

Cytogenetics Makes Its Contribution to Evolution

Mathematical population and breeding analysis played their role in shaping "the new synthesis," as the marriage of Darwinian evolution and classical genetics is sometimes called. An equally important role was played by cytogenetics. The first inroad was nondisjunction, noted in its consequence but not in its mechanism and relation to the expression of genes by R. Ruggles Gates (1882–1962) in *Oenothera*.[7] The elaborate use of nondisjunction to explain aneuploidy and how genes would be expressed with a chromosome loss or gain came from Calvin B. Bridges's extensive work correlating such sex chromosome gains or losses in a gamete and the white-eye mutation that would be anomalously expressed or hidden among the offspring.[8] If a female was white-eyed and she was mated with a fertile red-eyed male, the expectation in the F_1 generation was all white-eyed sons and all red-eyed daughters. But nondisjunction could lead to eggs carrying both X chromosomes, and fertilization by Y-bearing sperm resulted in white-eyed daughters that were predicted to be XXY and that, indeed, turned out to be so. Similarly, in that same cross, if a red-eyed male appeared in the F_1 generation, it would be predicted to be XO (lacking a Y chromosome) and hence sterile. That indeed was so. Bridges's work was admired also because the entire percentages of different aneuploid classes of offspring could be accounted for and predicted from each cross involving a euploid or aneuploid fertile parent. Bridges soon showed that aneuploidy for the smallest fourth chromosome was possible, but the offspring were severely impaired. This was also true for the triplo-X females, which were both sterile and physiologically frailer.[9]

The fly group argued that in animals aneuploidy in the autosomes was usually incompatible with embryonic development and thus did not survive. Sex chromosome aneuploidy complicated survival for the next generation because of the sterility of many aneuploid sex chromosome constitutions. But some plants were more tolerant of aneuploidy and could handle a third autosome or even some with a hemizygous autosome. Alfred F. Blakeslee studied this in detail among the Jimson weeds (*Datura*) and showed that these could persist in the wild.[10] Many domesticated strains of wheat and other plants had aneuploid varieties. Less damaging to development, however, was the presence of whole sets of chromosomes resulting in triploids, tetraploids, and even octaploids among some species of flowering plants. The absence of sex chromosomes made sterility less of an issue, but the pairing and meiotic separation of polyploid homologous chromosomes could result in problems, and this could reduce fertility by aborting many of the seeds.

Besides whole chromosome abnormalities that had detrimental consequences on species, there were a number of chromosomal arrangements whose evolutionary significance could be significant. Theodosius Dobzhansky (1900–1975) and Alfred Sturtevant studied chromosome inversions in natural populations of *D. melanogaster* or in other species of *Drosophila* or among the several genera of the Drosophilidae. The inversions contributed to the formation of what evolutionary biologists called "isolating mechanisms" that promoted differentiation of the separated populations into different species that would no longer breed with one another.[11] J.T. Patterson (1878–1960) and Wilson Stone at the University of Texas performed similar and more extensive studies in the American Southwest and Mexico in the late 1930s and 1940s. These more detailed studies included the new salivary chromosome techniques to identify breakpoints in the inverted chromosomes.[12]

A major contributor to an understanding of the evolution of chromosomes came from the work of Barbara McClintock. She worked out the mechanisms of recombination in ring chromosomes, chromosome inversions, and translocations.[13] She demonstrated why some plants carrying such rearrangements of chromosomes were infertile or why some classes of the recombination process were lethal to the embryo and aborted development. Much of that had to do with the aneuploid nature of crossover products or to aneuploid arrangements distributed to sperm or eggs. Equally important were the papers of Muller and his students in the USSR and in Edinburgh as they studied the evolutionary fate of inversions, translocations, deletions, and duplications.[14] They could account for the morphological changes in chromosomes such as converting a rod-shaped into a V-shaped or J-shaped chromosome or how two rods could form a V-shaped chromosome. They explained how a chromosome number could be decreased without expressing any of the ill effects associated with loss of large numbers of genes. This was particularly true for rearrangements involving two chromosomes whose reciprocal product consisted almost entirely of heterochromatic material. Such chromosomes could get lost with no noticeable damage to the offspring. Theodosius Dobzhansky's 1937 classic *Genetics and the Origin of Species* summarized much of the material that was emerging in numerous laboratories exploring many different species.[15]

Muller Offers an Evolutionary Interpretation of the Prevalence of Sexual Species

Muller would have been surprised at the phrase "Muller's ratchet" and the attention it has received from evolutionary biologists and geneticists. He did not identify it as his major contribution to genetics, and none of his other findings have the possessive of his name attached to his finding. The term was coined by Joe Felsenstein (1942–) in 1974 in an article on "The Evolutionary Advantage of Recombination."[16] Felsenstein drew on two articles by Muller. The earlier one, in 1932, discussed "Some Genetic Aspects of Sex."[17] The other, in 1964, discussed "The Relation of Recombination to Mutational Advance."[18] In the 1932 article, Muller argued that genetics had solved the "age old problem of the reason for the existence . . . of sex. . . ." Sex, he claimed, is not a necessity but a luxury, allowing more efficient competition with asexual species. Inbreeding in sexual species leads to a similar slowing down of variation in its introduction into competition. The price paid for vigorous sexual recombination is the elimination of so many more progeny per generation than in species that breed by asexual means. The asexual strains will depend on the compensation of heterozygosity for most new mutations that arise. But eventually this will lead to a deterioration of the diploid parthenogenetic lines. Muller did not use the term "ratchet" in the 1932 paper, but he did so in 1964, when he presented the process of asexual accumulation of deleterious mutations in asexual species or in inversion heterozygotes to a ratcheting of mutational load each generation.

Much of the debate on Muller's ratchet concerned asexual species like bdelloid rotifers, which have lacked sexual reproduction (as well as males) for at least 50 million years.[19] These species use three different ways to escape the deterioration of asexual reproduction. The first is their history of a past series of entire genome copying resulting in a tetraploid chromosomal composition. The second is the use of repair enzymes to generate a class of mutations called "gene conversions" that can occur without a sexual means of their formation. Rotifers are extremely resistant to radiation damage because of this process. The third is "horizontal gene transfer" from ingested DNA of other rotifers as well as occasional genes from other species. The inserted elements, when adaptive, persist and prevent Muller's ratchet, that is, the irreversible accumulation of deleterious mutations, escalating to species extinction.

The study of evolution depends on fragments of the past for its data and a lot of suppositions to fill in the broader picture of how species evolve, how higher categories arose, how continuous or discontinuous speciation is, and whether the speed of change is relatively rapid (e.g., dozens of generations or less) or slow (e.g., thousands of generations or more). There are also debates about how much change can come about through natural selection and how much by chance drift through isolating mechanisms. Bateson's developmental anomalies, dismissed or ignored for almost a century, have come back in the field of evo-devo, where such changes in segmentation, placement, and patterning play a major role in the origin of higher categories and where species differentiations within a genus are considered compatible with classical Darwinian selection, sometimes designated as "microevolution."[20] This makes evolution theory tentative and subject to change as more data enter the field in the form of new fossils or nucleotide sequences from contemporary and more removed ancestry.

Molecular Biology and Genomic Approaches Contribute to Understanding Evolution

The development of molecular biology had a major influence on evolution studies because each living organism is a palimpsest of sorts, with the present genome partially written over successive past genomes in phylogenetic pasts. This occurs at both the level of the gene and the level of the genome.[21] The genic studies came first with the study of hemoglobin (1950s to 1980s) in human populations, and then among the primates and other mammals, and then among all animals bearing hemoglobin, including sipunculoid worms. These studies showed that the two chains had separated in a distant past as α and β chains and that similar sequences were in hemoglobin chains associated with embryonic (Hb-ε) or fetal (Hb-F) respiration of oxygen. These studies also showed similarities of sequence in many of the molecules receiving oxygen from circulating red blood cells, including myoglobin in muscle cells. The studies in those mid-century years were confined to the sequences directly involved in making components of the two protein chains, one (Hb-β) a nest of genes on chromosome 16 in humans and the other (Hb-α) on chromosome 11. It was the β chain that housed the notorious allele change in the number 6 position that led to sickle cell anemia.

When genes were recognized as split into introns and exons, this opened up a study of the base-pair differences within the excised introns. They could serve as "genetic markers" or "single nucleotide polymorphisms" (SNPs) to distinguish populations from one another in a species or to measure the amount of time elapsing between changes across the species of cognate genera, or higher taxonomic categories. This set up the concept of using mutational differences in introns as a "molecular clock" to determine when different evolutionary groups separated from each other.[22] If two lines shared a common set of markers, then those had to have existed before the split. If the markers differed in two groups, then independent mutational histories were evident in that region of the gene. Unlike the changes in exons, which are directly related to the pressures of natural selection, the base changes in the introns are mostly what can be called "neutral mutations" playing no significant role in the fate of the organisms bearing them. Like a palimpsest, however, they can reveal their ancestral origin.

As the capacity to sequence DNA improved with new technologies in the late 1990s and early 21st century, the number of genes that could be deciphered at a reasonable cost rose dramatically, and whole-genome sequencing became possible first for viruses and bacteria and then for eukaryotic cells. The Human Genome Project was effective and endorsed by most geneticists because it led to the genomic sequencing of virtually every widely used laboratory organism, and many domesticated plants and animals. For evolutionary biology, this was a glorious time because "comparative genomics" using these sequences showed when divergences took place, when duplications of the genome occurred, and how genes moved about within and across different branches of the phylogenetic tree. Molecular biologists learned to use predicted amino acid sequences to identify domains of genes associated with specific functions in different tissues, which allowed genome analyses that classified unique genes and shared genes, genes associated with cell functions, and genes associated with organ functions. So far, as the first decade of the 21st century ended, the percentage of sequenced genes in a eukaryotic organism far outnumbers the assigned functions to the genes in a genome.

The molecular analysis of genomes reveals that the farther apart two species are, the more their conserved genes are shifted about to different chromosomes. Humans and chimpanzees are close species that separated about 5 million years ago. They differ by nine major inversions, one translocation (forming human chromosome 2 from chimp chromosomes 2A and 2B) that reduced the chromosome number from 48 to 46 with some loss of heterochromatin, and numerous (about 5 million) minor rearrangements, especially tandem duplications, minute deletions, and insertions.[23] At the nucleotide level, there are about 35 million SNPs. This still leaves 30% of the genes in both species identical in sequence. What is particularly satisfying is the agreement between cytological, anatomical, and genomic approaches to generating lines of descent among cognate and more distant ancestors.

One additional innovation in the use of DNA sequencing has been the successes in extracting DNA from mummified or partially fossilized remains of extinct species. This allows a glimpse of past functions and an opportunity to make phylogenetic connections to living descendants.[24] Whether artificial nuclei can be reconstituted, transferred to a suitably close enucleated ovum, and function there remains to be seen. Such efforts are being contemplated for mammoths, with living elephants as the source of eggs deprived of their own nuclear DNA.

Michael Lynch (1951–) and his laboratory use a different approach with molecular tools to study microscopic organisms such as *Daphnia, Paramecia, Chlamydomonas,* and other organisms over thousands of generations.

They study whole-genome sequences to identify the mutation rates per generation, the spectrum of mutational lesions and developmental effects, and the roles of repair enzymes, recombination, and other factors involved with the evolution of mutation rates.[25]

Repair Enzymes Play a Role in Mutagenesis and Chromosomal Rearrangement

In 1944, while she was at Cold Spring Harbor, Evelyn Witkin (1921–) discovered a mutant strain of *Escherichia coli* that was resistant to ultraviolet radiation. This began a life-long interest regarding the mechanisms of mutation and the means by which cells, chromosomes, and genes can bring about repair from exposure to agents that alter them.[26] Repair enzymes are probably as old as the origin of the first cells. Damage to DNA is acceptable if two conditions are met. First is the need to repair most chromosomes that have breaks (single or double stranded) induced in them. Second is the need for some inefficiency in base-pair replacement that allows point mutations to occur. If the process were perfect, mutation rates might have been too low for the rapid evolution of new species or adaptations needed for abrupt environmental changes. Repair enzymes are thus important in generating evolutionary change.

In 1944, genes were seen as nucleoproteins, with some uncertainty about the roles of the two chemical components composing them. The work of Oswald Avery, Colin MacLeod, and Maclyn McCarty settled the issue, and after 1945, Witkin and most geneticists using microbial systems shifted to an interest in DNA as the major constituent of the genes found in eukaryotic chromosomes. Also in the early 1940s, both Barbara McClintock, then at the University of Missouri, and Hermann Joseph Muller, then at the University of Edinburgh, recognized a type of repair process involved in the repair of spontaneous or induced chromosome breakage. For McClintock, *Zea mays* was more complex than Muller's fruit flies because such breaks had two options. Some healed and the broken end acted like a new telomere or chromosome tip with just a simple loss of the genes distal to the break. Most, however, joined another broken end with some sort of chromosome rearrangement as the outcome. This could lead to the production of the long-known deletions, duplications, inversions, or translations that had been found in plants and animals for most of the first half of the 20th century. But McClintock, and independently, Muller and Guido Pontecorvo, found a category of rearrangement in which a broken chromosome replicates and joins to its replicated component, forming aneucentric chromosomes with no centromere or with two centromeres. If the centromeres are far enough apart, they form a chromosome bridge during mitotic or meiotic anaphase, and this leads to a cycle of further breakage and reunion, resulting in aneuploid cells. The fruit fly embryo usually aborts as a dominant lethal when such aneucentric chromosomes form from breakage of a sperm or egg chromosome. Animals in general do not tolerate the accumulating aneuploidy or the altered timing of cell division. Plants, however, may lose portions of their genetic material in such cell lines, producing spotting and streaking of kernels in maize, a typical response that McClintock observed and interpreted in her maize plants. She named the process the breakage–fusion–bridge cycle.

Since Witkin's discovery of a repair mechanism for DNA in ultraviolet exposure, many types of repair enzymes have been isolated. Human cells experience about 1 million lesions per day per cell. The causes include external agents like ultraviolet light from the sun; X rays or other ionizing radiation from medicine, industry, or natural background radiation; hydrolysis of nitrogenous bases from heat; toxins that affect DNA; chemical mutagens, especially peroxides, many of them associated with cellular metabolism; and viruses whose DNA may enter and leave human chromosomes and in the process take out or leave behind small amounts of DNA.[27] These agents may act in different ways. Some cause errors occurring during the transcription process when DNA is copied into messenger RNA. Some cause lesions occurring in exons, which would have a higher chance of producing mutational damage than lesions occurring in the introns. Some cause single-strand breaks, which are likely to interfere with transcription of DNA to messenger RNA. Some cause two-strand breaks, which lead to chromosomal rearrangements if they are not immediately restored to their original configuration and association. Some cause cross-linking events such as thymine dimers that form when DNA absorbs ultraviolet radiation. The cross-linking of DNA has consequences similar to McClintock's breakage–fusion–bridge cycle.

Cells respond in different ways to chromosome breakage and repair. Some become senescent, and their efficiency diminishes over time. Some trigger an event called apoptosis, in which the cell self-destructs.[28] From a human perspective, the most unwanted form of response is tumor cell formation when the cell enters into unregulated cell divisions, often with loss of tissue recognition and loss of adhesiveness resulting in metastatic cancer.

In the process of excising defective bases, nucleotides, cross-linkage, or mismatched pairing as a consequence of the induction of mutations after

DNA damage, transitions and transversions might occur as simple base substitutions. Sometimes losses of nucleotides result in frameshift mutations, and these can be damaging to the gene, depending on where that lesion had taken place along its length. The joining of broken ends in two-strand breakage can lead to isolating mechanisms in evolution. Inversions and translations often prevent or reduce fertility between cognate populations, as de Vries eventually discovered in his *Oenothera* species and varieties.

Mutations of different repair enzymes in humans result in some very serious genetic disorders, including Cockayne syndrome, xeroderma pigmentosum, Bloom syndrome, Fanconi anemia, and ataxia telangiectasia.[29] These vary in severity, but most are associated with elevated cancer frequencies, stunted growth, reduced life expectancy, and multiple organ involvement. What these clinical disorders reveal is a dependency in eukaryotic cells on numerous repair enzymes for different types of lesions in DNA resulting in mutations, single-stranded breaks, two-stranded breaks, cross-linking of DNA, and other defects. In eukaryotic cells, repair enzymes move along the DNA and identify specific type of defects for which they make a repair. The high induced mutation rates associated with the repair process contribute to gene evolution at the same time as the repair enzymes conserve the integrity of the chromosomes and the DNA for performing its essential roles of replication, transcription, and RNA processing.

Mitochondrial DNA Differs from Nuclear DNA in Eukaryotic Cells

A considerable amount of research on evolution is devoted to mitochondrial DNA. Mitochondria are organelles found in the cytoplasm of eukaryotic cells. Their major function is oxidative metabolism, a process by which adenosine triphosphate (ATP) is produced, combining oxygen and small carbon compounds using enzymes in the inner membrane of the mitochondrion. The number of mitochondria per cell varies with the tissue in humans, but 10 or more is not uncommon. In addition, mitochondria have their own DNA. The DNA in human mitochondria has 16,569 base pairs. The DNA is circular, and there are usually multiple copies of the DNA in the matrix of the mitochondrion. The genome of the mitochondrion produces 37 genes in humans, most of them associated with the proteins and the RNA involved in oxidative phosphorylation. The proteins of the membranes of mitochondria have a dual origin, some produced by the nuclear genes and others by the mitochondrial DNA.

Mitochondria are of evolutionary interest for two major reasons. Lynn Margulis popularized an idea that was current in the early 20th century that mitochondria arose from a process called endosymbiosis.[30] This was stimulated by the observations in the 18th and 19th centuries that some organisms, such as species of hydra, harbor symbiotic cells (algae) that perform photosynthesis.[31] In *Paramecia,* one species, *Paramecium bursaria,* has similar zoochlorellae, as they are called, present that act like chloroplasts. Margulis and her students showed that there was a similarity between the mitochondrial DNA and proteins to bacteria. She believed that the origin of eukaryotic cells might have involved a mechanism in which such symbiotic inclusions transferred or lost genes and were selected for a primary function such as oxidative phosphorylation in animal cells and for chloroplast photosynthesis in plant cells. Both animal and plant cells would have mitochondria, but chloroplasts would be selected in addition in plant cells. There is strong support for the endosymbiosis model of mitochondria and chloroplasts. The genes are relatively small in number compared to those of the bacteria that have an oxidative metabolism or bacteria that act as blue-green algae with a capacity for photosynthesis.

One feature of mitochondrial inheritance in animals is that it is primarily transmitted through the female's eggs and rarely, if at all, from the sperm. In animal sperm, the mitochondria are tightly compacted in the midpiece of the mature sperm and assist in the contraction of the tails when sperm are activated after ejaculation. The midpiece does not enter the egg, or if it does, its mitochondrial mass is enzymatically degraded. There may be mutations that permit occasional survival of sperm mitochondria or that fail to digest the relatively infrequently transmitted midpiece. Overwhelmingly, efforts to find such transmission by radioactive labeling or genetic markers have shown that the process, if real, is rare. This transmission through the egg leads to a condition called maternal inheritance. Some human genetic disorders are associated with mutations in mitochondrial DNA. Leber optic atrophy results in foveal degeneration of cone cells leading to loss of color vision and a scotoma or blind area associated with massive loss of those cells.[32] The condition can spread peripherally in the retina and lead to total blindness. Males with Leber optic atrophy do not transmit their defect to their offspring.

Complicating genetic studies of mitochondrial DNA are biological aspects not encountered with nuclear DNA. Usually a person with a human mitochondrial disorder has cells showing heteroplasmy in which there are some mitochondria that have the mutant DNA and others that do not.

This can make the expression of the disorder variable among siblings from a female whose mitochondria are affected by a newly arising mutation in her germline or in some earlier ancestor. Also complicating the study of mitochondrial DNA is the much higher mutation rate found in mitochondrial genes than in nuclear genes. This is due to two factors. The mitochondrial DNA lacks repair enzymes, and the process of oxidative phosphorylation leads to a higher oxygen-induced mutation rate of the DNA. There are also no repetitive regions of DNA, and there are no introns in human mitochondrial DNA.

Despite these differences, attempts have been made to follow mitochondria in populations within a species, such as the history of human migrations out of Africa to other parts of the world, and among species, to show how species within a genus have diverged. Allan Wilson (1934–1991) and his students were early contributors to these two approaches.[33] In popularized accounts, the human transmission out of Africa is given the name "mitochondrial Eve," which to geneticists includes cautionary thoughts that not one female but many females with the same genomic composition of their mitochondria might have emerged from Africa or that many types might have moved out, but only one or a few major types prevailed at the time racial differentiation began to take place. For this reason, evolutionary biologists have used additional approaches, including Y-chromosome transmission in human populations (known to the non-scientist reader as "Y-chromosome Adam") and nuclear gene transmission from ethnic groups to look for correlations and a consistent trail of human genes as the earth became populated with our species. Unfortunately, the use of these popular terms to introduce mitochondrial and Y-chromosome anthropological and evolutionary work in humans has led to an immense interest by creationists, who oppose an evolutionary history of humans and who have made efforts to shift those terms into a biblical time span of about 6000 years.

The changes in the idea of mutation in relation to evolution have been numerous, and they reflect the way each generation sees a piece of an emerging picture. For Darwin's day, that was limited largely to fieldwork, domesticated varieties, and comparative anatomy. A generation later, it shifted to mathematical approaches that frustrated Bateson, de Vries, and Morgan, who sought more experimental approaches. The third generation introduced classical genetics; the fourth generation introduced the new systematics; the fifth generation introduced biochemical approaches, especially with sickle cell anemia and its relation to the hemoglobin model; the sixth generation introduced molecular approaches and comparisons of genomes among cognate species; and the current generation is following leads from evo-devo developmental genetic models and epigenetic studies of how protein domains lead to multiple clusters of proteins. Each new addition of concepts, in turn, has been driven by new technologies that allowed evolution to probe in new directions not available or not conceivable to earlier generations.

References and Notes

1. Cleland R. 1932. Further data upon circle formation in *Oenothera*, its cause and genetical effect. *Genetics* **17:** 572–602.
2. Hardy GH. 1908. Mendelian proportions in a mixed population. *Science* **28:** 49–50. Weinberg W. 1908. Über den Nachweis der Vererbung beim Menschen. *Jahreslerte des Vereins fur Vaterlandische Naturkunde im Würtemberg* **64:** 369–382.
3. Danforth C. 1923. The frequency of mutation and the incidence of hereditary traits in man. In *Eugenics, genetics, and the family,* Vol. 1, pp. 120–128. Scientific Papers of the Second International Congress of Eugenics Held at the American Museum of Natural History, New York, September 22–28, 1921. Williams and Wilkins, Baltimore.
4. Muller HJ. 1950. Our load of mutations. *Am J Hum Genet* **2:** 111–176.
5. Provine W. 1971. *The origins of theoretical population genetics.* University of Chicago Press, Chicago.
6. Based on discussions with Krishna Dronamraju, a student of Haldane's who told me of the friendship between Haldane and Mayr. See Dronamraju K. 2011. *Haldane, Mayr and beanbag genetics.* Oxford University Press, New York.
7. Gates RR. 1958. *Taxonomy and genetics of* Oenothera*: Forty years study in the cytology and evolution of the Onagraceae.* Springer, New York.
8. Bridges CB. 1916. Non-disjunction as proof of the chromosome theory of heredity. *Genetics* **1:** 1–52; 107–163.
9. Bridges CB. 1913. Non-disjunction of the sex chromosomes of *Drosophila. J Exp Zool* **15:** 587–606.
10. Blakeslee AF. 1931. *Extra chromosomes, a source of variations in the Jimson weed.* Annual Reports of the Smithsonian Institution 1930. US Government Printing Office, Washington, DC.
11. Dobzhansky T, Sturtevant AH. 1938. Inversions in the chromosomes of *Drosophila pseudoobscura. Genetics* **23:** 28–64.
12. Patterson JT, Stone WS. 1952. *Evolution in the genus* Drosophila. Macmillan, New York.
13. Rhoades MM, McClintock B. 1935. The cytogenetics of maize. *Bot Rev* **1:** 292–325.

14. Muller HJ. 1938. The remaking of chromosomes. *Collect Net* **13:** 181–198. Muller HJ. 1940. Bearings of the *Drosophila* work on systematics. In *The new systematics* (ed. Huxley J), pp. 185–268. Clarendon Press, Oxford.
15. Dobzhansky T. 1937. *Genetics and the origin of species.* Columbia University Press, New York.
16. Felsenstein J. 1974. The evolutionary advantage of recombination. *Genetics* **78:** 737–756.
17. Muller HJ. 1932. Some genetic aspects of sex. *Am Nat* **66:** 118–138.
18. Muller HJ. 1964. The relation of recombination to mutational advance. *Mutat Res* **106:** 2–9.
19. Mark Welch DB, Mark Welch JL, Meselson M. 2008. Evidence for degenerate tetraploidy in bdelloid rotifers. *Proc Natl Acad Sci* **105:** 5145–5149.
20. Field KG, Olsen GJ, Lane DJ, Giovannoni SJ, Ghiselin MT, Raff EC, Pace NR, Raff RA. 1988. Molecular phylogeny of the animal kingdom. *Science* **239:** 748–753.
21. Bouriat S, Nielsen C, Economou AJ, Telford MJ. 2008. Testing the new animal phylogeny: A phylum level molecular analysis of the animal kingdom. *Mol Phylogenet Evol* **49:** 23–31.
22. Zuckerkandl E, Pauling L. 1962. Molecular disease, evolution, and genetic heterogeneity. In *Horizons in biochemistry* (ed. Kasha M, Pullman B), pp. 189–225. Academic Press, New York. Sarich V, Wilson AC. 1967. Immunological time scale at the molecular level. *Nature* **217:** 634–626.
23. Caswell JL, Mallick S, Richter DJ, Neubauer J, Schirmer C, Gnerre S, Reich D. 2008. Analysis of chimpanzee history based on genome sequence alignments. *PLoS Genet* **4:** e1000057. doi: 10.1371/journal.pgen.100057.
24. Handt O, Krings M, Ward RH, Pääbo S. 1996. The retrieval of ancient human DNA sequences. *Am J Hum Genet* **59:** 368–376. Miller W, Drautz DI, Ratan A, Pusey B, Qi J, Lesk AM, Tomsho LP, Packard MD, Zhao F, Sher A, et al. 2008. Sequencing the nuclear genome of the extinct mammoth. *Nature* **456:** 387–390. Dalton R. 2010. Ancient DNA set to rewrite human history. *Nature* **465:** 148–149.
25. Lynch M. 2010. Evolution of the mutation rate. *Trends Genet* **26:** 345–352.
26. Witkin E. 1960. Ultraviolet induced mutation and DNA repair. *Ann Rev Genet* **3:** 525–552.
27. Friedberg C, Walker GC, Siede JW. 1995. *DNA repair and mutagenesis.* American Society of Microbiology, Washington, DC.
28. Kerr JF, Wylie AH, Currie AR. 1972. Apoptosis: A basic biological phenomenon with wide-ranging implications in tissue kinetics. *Br J Cancer* **26:** 239–257.
29. Bridges BA. 1981. Some DNA-repair-deficient human syndromes and their implications for human health. *Proc R Soc Lond B Biol Sci* **212:** 263–278.
30. Sagan L. 1967. On the origin of mitosing cells. *J Theor Biol* **14:** 255–274. Lynn Margulis published under her married name, Sagan. After her divorce from astronomer Carl Sagan, she published under her maiden name, Margulis.

31. Lenhoff SG, Lenhoff HM, eds., trs. 1986. *Hydra and the birth of experimental biology, 1744: Abraham Trembley's memoires concerning the Polypes.* Boxwood Press, Pacific Grove, CA.
32. Wallace DC, Singh G, Lott MT, Hodge JA, Schurr TG, Lezza AM, Elsas LJ II, Nikoskelainen EK. 1988. Mitochondrial DNA mutation associated with Leber's hereditary optic neuropathy. *Science* **242:** 1427–1430.
33. Cann RL, Stoneking M, Wilson AC. 1987. Mitochondrial DNA and human evolution. *Nature* **325:** 31–36.

9

Mutation in Relation to Genetic Engineering

PLANT AND ANIMAL BREEDERS IN THE EARLY DECADES of the 20th century did not consider themselves genetic engineers. They were limited mostly to what nature provided, and only after Hermann Joseph Muller's announcement of induced mutations with X rays did the possibility arise that mutagenesis was not only a branch of genetics but a tool for applied science. X rays were used to produce new breeds of cereal grains in Sweden, to boost the yield of penicillin from strains of the fungus producing this antibiotic, and to work out biochemical pathways that might have useful applications in the pharmaceutical industry. All of these applications still used the same skills of the breeder—devising stocks or conditions to isolate desired mutations and perpetuate them by using recombination to insert them into a desired genotype.

The term "genetic engineering" became more widespread when the tools of genetics changed dramatically after 1970 as geneticists learned how to sequence genes, how to cut chromosomes into small bits and isolate desired genes, and how to construct genes to order, replacing any desired base pair to produce what was called "directed mutation." When geneticists first realized they could use restriction enzymes to cut any DNA of any species into small bits that could be isolated and placed into viral or synthetic "vectors," they knew that horizontal transfer of genes from one species into another was no longer a rare accident but a technical possibility.[1] They could put genes for making fluorescent pigments like those in fireflies into vertebrates and make them glow in the dark. But human imagination is not limited to pleasing novelties that make good images for popular publications. It is also applied to matters of health, including research in genetic disorders or targeting cancer cells or other diseased tissues, such as the

TABLE 1

Chronology of main events relating to development of methods for constructing and cloning recombinant DNAs

Year[a]	Event
1969–1970	P. BERG (1970) and LOBBAN (1969) independently conceive ideas for generating recombinant DNAs *in vitro* and using them for cloning, propagating, and expressing genes across species.
1971	D. BERG *et al.* (1974) isolate the first plasmid bacterial cloning vector, λ*dvgal* 120.
1971	Concern regarding potential biohazards of cloning first raised by Robert Pollack.
1971–1972	JACKSON *et al.* (1972) and LOBBAN and KAISER (LOBBAN 1972; LOBBAN and KAISER 1973) concurrently and collaboratively develop the terminal transferase tailing method for joining together DNAs *in vitro*.
1972	JACKSON *et al.* (1972) create first chimeric DNA *in vitro*.
1972	MERTZ and DAVIS (1972) discover that cleavage with *Eco*RI generates cohesive ends. They use *Eco*RI plus DNA ligase to generate SV40-λ*dvgal* 120 chimeric DNAs *in vitro*.
1972–1973	COHEN *et al.* (1972) isolate the drug-selectable bacterial cloning vector, pSC101. They use it to construct, clone, and express bacterial intra- (1973) and interspecies (1974) recombinant DNAs.
1973	MORROW *et al.* (1974) clone and propagate ribosomal DNA genes from a eukaryote in *E. coli*.
1973–1976	Renewed concerns regarding potential biohazards of cloning recombinant DNAs (SINGER and SÖLL 1973; P. BERG *et al.* 1974, 1975) lead to NIH *Guidelines*.
1974–1975	Filing of initial Stanford University/University of California, San Francisco (UCSF) (Cohen/Boyer) patent applications relating to recombinant DNA.
1976	Boyer and Robert Swanson cofound Genentech, the first biotechnology company.
1980	Stanford/UCSF (Cohen/Boyer) patent issued by U. S. Patent Office.

[a]Year(s) in which event occurred.

Paul Berg, Herbert Boyer, and Stanley Cohen develop recombinant DNA technology. The idea of using restriction enzymes for genetic engineering began in 1969. By 1980 the idea shifted out of academic interest and into commercial companies and the awarding of patents. This is Berg's recollection of the events. (Reprinted, with permission, from Berg P, Mertz JE. 2010. Personal reflections on the origins and emergence of recombinant DNA technology. Genetics 184: *9–17. © Genetics Society of America.)*

plaque-forming regions of the coronary vessels of the heart or the neuronal tangles and clumped proteins that are associated with dementia—desired applications for most of humanity. Human imagination becomes compromised when technology is turned to bringing death, injury, or other harm to others. Bad outcomes arise when genes are involved in extending host ranges of pathogens. They arise when microbes are altered to make the toxins more potent or in greater concentration for use in germ warfare. They arise, without any ill intent, if a scientist accidentally releases a highly infectious altered disease organism that was being studied to help humanity such as reconstructing the flu virus that caused the 1919 epidemic that killed tens of millions of young people around the world.[2]

Many prominent geneticists who were aware of this new application of genetics, called "recombinant DNA technology," initially worked out by Herbert Boyer, Stanley Cohen (1922–), Paul Berg, and others, were concerned about abuse and accident. Some geneticists and critics of recombinant DNA technology feared unintended mutations occurring in the recipient altered bacteria or the deliberate insertion by vectors of genes that would be deleterious to human health. Was this an unjustified fear? To assess that concern, they met several times at conferences to express their worries and to set up guidelines that would keep the inexperienced from making careless uninformed experiments and to assure themselves and the public that stringent safety measures would be used when working with recombinant DNA technology, at least until the field was assessed as safe.[3] This led to self-regulation in industry and a quasi-governmental regulation by requiring grant recipients to abide by those standards of safety.

Most scientists are motivated by the benefits of their applied work, and most pharmaceutical companies are motivated by the good their products can offer to those who are ill and the profits they can make from good products. The first wave of products that came out of the recombinant DNA technology was much-needed products like human insulin and human growth hormone.[4] Before this new technology, insulin was derived from the pancreases of animals in slaughterhouses. Human growth hormone was obtained from human cadavers. Both were unsatisfactory sources. The pig insulin in some diabetics acted as an antigen. The human growth hormone was sometimes contaminated with agents that cause neurological degeneration, like Creutzfeldt–Jacob disease. The concern with germ warfare was addressed by an international agreement to ban chemical and biological warfare and to close those facilities that did such research and destroy the stockpiles of agents already manufactured for war use.

More than a generation has passed since those first efforts at using recombinant DNA technology, and most of the restrictions of the early years have been removed. Highly modified strains of *Escherichia coli*, the microbe most often used in producing large quantities of a desired product from an inserted gene, have never accidentally produced a pathogenic form of *E. coli* capable of living outside the laboratory. Some U.S. scientists resented any regulation, even by their peers, but most recognized that they would be considered like physicians, who are highly regulated, and the drug industry, which has to abide by the standards set by the Food and Drug Administration.[5]

The tools for genetic engineering have proliferated. A desired gene or mutation can be cloned in large quantities by using the technique thought out by Kary Mullis (1944–) and colleagues at Cetus, a company largely devoted to recombinant DNA technology and its products.[6] He used a heat-resistant enzyme derived from *Thermophilus aquaticus* (designated as *Taq*) to keep the replication process going with his primer-marked regions of DNA that were being replicated in a chain reaction process. He called this a "polymerase chain reaction" (PCR), and it rapidly entered research and commercial usage. PCR can be used for cloning genes or pieces of desired DNA. It is also used in forensics to amplify DNA found in crime scenes, and that has supplemented or replaced fingerprint evidence as a basis of convicting or exonerating accused individuals during crime investigations. It also entered public awareness of science by its use in many television crime dramas. It can also be used to identify the DNA of historical figures or mummified tissues of partially fossilized human ancestors and other species.

For geneticists, the tools of genetic engineering have opened up immense areas of research. Comparative genomics allows comparisons of all forms of life, and a project is under way to have at least one representative of each of the 92 known phyla sequenced so they can be used to augment the morphological and classical techniques for taxonomic classification. Complete genomes allow unique features of each phylum to be identified at a molecular level. They also allow, within a particular species, access to genes set aside for cell functions, developmental events in the life cycle from fertilization to death, and organ-specific genes that are associated with both the organ's assemblage and its functioning. Living things have a history of at least 3.5 billion years since prokaryotes abounded on earth. With so many generations (most of them relatively short compared to humans), spontaneous mutations, recombinations, rearrangements of the DNA, and natural selection sifting all these mutational differences, the ancestry of any living thing today is historically complex. Working out that complexity may take many generations of scientists to come.

Applications of recombinant DNA technology shifted in public attention from initial fears of epidemics (1970s), to concerns a generation later regarding using the technology for human products (insulin, other hormones, clotting factors) as possibly being abused (e.g., cheating in sports by use of human growth hormone), and to concerns regarding eating or making genetically modified foods. All new technologies dealing with the human body have encountered controversy when they were first introduced. This is true for vaccination to prevent infectious diseases, fluoridation of water supplies to prevent dental caries in children, chlorination of water supplies, use of vitamin supplements, and many other measures that over time have proven largely to be safe and beneficial. Abuse can always exist when products are used improperly. On some occasions, errors will be made if regulation is not a common practice. The sale of sulfa drugs in a toxic solvent in the 1930s caused many deaths and led to a very stringent Food and Drug Administration in the United States, which did its job and prevented the importation of thalidomide for licensing because Frances Kelsey (1914–) did not like the faulty testing in Europe.

Genetically modified foods are based on the idea that genes can be isolated to improve the commercial qualities of food crops.[7] Farmers would plant wheat that was resistant to fungal infection or seeds that would provide plants that did not snap or uproot in the presence of heavy winds. They would welcome growing crops like rice in areas where the water has more salt content than is normally tolerated by rice plants. They would consider it a benefit to have drought-resistant strains to plant in relatively arid regions. Supermarkets would love to extend shelf life, improve flavor, provide an optimum texture, maintain a pleasing color, and sell varieties that are more nutritious to the families eating them. World health organizations would love to see the development of rice, wheat, maize, mullet, and other cereal grains that have amino acid enrichment to make them suitable for customers with primarily vegetarian diets, that would boost vitamin content, and that might even have oral agents that prevent human infectious diseases.

Those are the incentives that stimulate scientists to insert directed mutations, trans-species genes, or someday completely designed genes that provide desired characteristics to such foods. Opposed to all of the attempts to design genetically modified foods are traditionalists who believe that only organic foods are suitable for their consumption and who would not enter most supermarkets because they reject any crops using chemical agents to combat plant diseases. There will always be a small percent of the food market dedicated to organic food stores for those rejecting any artificial processes or supplements in the making of their foods. A much larger percent of humanity is frightened by stories of potential harm from introduced gene products acting as allergens. If the allergens found in peanuts, for example, were somehow introduced into rice, wheat, or maize, they would be life-threatening for such sensitive individuals if they were not warned of that risk. Some argue that, although they would avoid such products if they were properly labeled, they want to be assured that the genes in such crops will not find their way by accidental pollination into crops that were thought to be allergen-free. In a highly diverse human population in industrial counties with many consumers and many options to enlarge the varieties of foods they eat, this is a problem that may not be totally preventable, and wherever such low-risk harm occurs, it is often measured against benefits that outweigh the risk of harm. The burden of navigating through foods in such a world of genetically modified foodstuffs is likely to be put on the persons who are sensitive. If they are well educated, they could limit their diets to trusted suppliers and trusted foods. But most people are not well educated, and that is where society has not resolved how to deal with low risks. Despite those controversies, the field of genetically modified foods using recombinant DNA technology to identify and to insert desired mutations is a growing one and in the generations to come will very likely account for almost all commercially raised and sold foods in the world.

Another genetic tool for studying the genome and especially its application to genes associated with human disease is the development of "knockout mutations," especially in mice. Genes can be rendered nonfunctional by taking a sequence of DNA containing the signatures of a gene (its promoter and other reading sites and its termination site), adding alterations of one or more of its exons, and then inserting that into a blastomere (embryonic stem cell) and eventually obtaining a chimeric mouse that can be bred and a homozygous form isolated. The gene may abort the embryo or malform the mouse that is born or result in some impairment in the adult mouse. All of the genes in a mouse can thus be tested for their functions in the life cycle of mice.[8]

The Introduction of Genetic Services to Health and Medicine Gain Rapid Acceptance

Virtually the reverse concerns are found in the field of "genetic services." Instead of new products that frighten a segment of the population, the consumers for this field are those facing problems of infertility, having pregnancies at risk for birth defects, or trying to inform relatives of risks they can identify and prevent.[9] Infertility mutations in humans rely on a long history of attempts to understand why about 10% of humanity has difficulty having children and why a substantial portion of the fertile population has difficulty trying to carry out family planning by using measures to prevent fertilization. In fruit flies, X-linked mutations that affect fertility are about as common as those that produce X-linked lethals. Such fertility mutations affect gamete production, gamete function, or the earliest stages of embryonic development. In humans, infertility was untreatable except for a very few couples until the 1960s. By that time, those studying the field of reproductive endocrinology began to understand the role of hormones in spermatogenesis and oogenesis. The hormones controlling the menstrual cycle were also worked out. If one went back to the 1920s and 1930s, the treatments for infertility were indirect. There was a Kaplan treatment that applied about 200 roentgens to the pituitary gland and to the ovaries in the expectation that radiation stimulated hormonal activity, and this led to more pregnancies.[10] This was shown to be a placebo effect in the 1950s.[11] Muller was horrified by the procedure, especially the deliberate X-raying of ovaries at a relatively high dose for humans. Ira Kaplan (1887–1963), who had been doing this for many years, scoffed at the hazards of radiation and even cited second-generation patients whose mothers had come to him because they were sterile.

In the 1950s, some women benefited when sex hormones became available for therapeutic use, and many women could respond to the steroid hormones and ovulate. This sometimes had unintended consequences because women varied genetically in their responses. A dose that in most women would lead to the maturing of one egg in a menstrual cycle could lead in another woman to two or more eggs being released at once, and this produced lots of twinning. Contrary to the popular belief up to that time that assigned infertility to women, there was abundant evidence that males had almost as many infertility problems as females. This led to the testing of both partners seeking advice on their infertility. For males, sperm counts, sperm morphology, and sperm motility became routine parts of the study of a couple. By the 1980s, genes involved in the formation of sperm, their ability to migrate, their morphology, and their immunological recognition of the egg surface were identified.[12] Many of these turned out to be deletions or point mutations of genes in the Y chromosome. The Y chromosome was identified as having genes associated with male determination in humans, especially the *SRY* gene.[13] Loss of function of that gene could lead to "sex reversals" in which females (usually lacking ovaries) were XY but otherwise appeared to be normal females at birth. The *SRY* gene could also be transferred to an X chromosome by deletion and a subsequent insertion into an X chromosome, and this could produce XX males who had testes and male genitalia but who produced no sperm. Most of the genes in the Y chromosome were involved in spermatogenesis, and those were the genes that created most instances of male infertility.

The problems were more numerous in females because female infertility could be assigned to egg formation, egg release, uterine lining receptivity, ovarian surface recognition of sperm proteins, or oviduct damage that prevented egg transport from the ovary to the uterus. Whereas some instances were nongenetic, caused by infection with *Chlamydia* or gonococcal infections, most were genetic; however, the distribution of genes for these conditions was more scattered and harder to identify than the sperm-related disorders. Nevertheless, enough was known by the 1970s to try the first artificially assisted fertilization in vitro, and a physician trained in Virginia at the Jones Institute successfully did that in Great Britain.[14] Soon after, an American child was produced by this procedure, which shifted its name in the popular press from "making test tube babies" to "in vitro fertilization," usually abbreviated as IVF. Although IVF technology rapidly flourished in the 1990s and IVF clinics appeared in almost all major cities and hospitals, the isolation of genes involved in fertility mutations took longer to identify.

A generation earlier, the fields of genetic counseling and medical genetics began to emerge. They sought to identify mutations in couples at risk. In the 1940s the options were limited.[15] There was adoption if there was a serious disorder like Huntington disease in the family. There was artificial insemination if a male knew he was at risk for passing on a disorder to a child (a young man whose father or mother had Huntington disease would be such a candidate). There was also vasectomy or tubal ligation if the couple decided not to have children. "Prenatal diagnosis" did not become a possibility until the 1960s. It arose after interest in chromosomal aneuploids intensified from the finding that Down syndrome was associated with a trisomy for chromosome 21.[16] By 1962, two more autosomal conditions, one later identified as trisomy 13 (originally Patau syndrome) and the other later identified as trisomy 18 (originally Edwards syndrome), were identified.

At first in the 1950s, amniocentesis was used to treat an Rh blood incompatibility in utero by transfusion of blood to the fetus before its mother's immune system would harm the fetus. With the knowledge that fetal cells might be isolated, cultured, and examined under the microscope for their chromosomes, physicians began looking for women at risk to see if the procedure would work. It did so with no major injuries to the embryo. The techniques improved dramatically when ultrasound was added to steer the needle. The field also took another leap into diagnostic efficiency when Tjorborn Caspersson in Sweden developed chromosome-banding techniques.[17] He used quinacrine to reveal a banding pattern in the metaphase chromosomes that allowed the identification of each of the chromosomes as well as rearrangements within them or between any two of them.

While human cytology and cytogenetics led the way, biochemists were learning how to identify the presence or absence of essential substances in the fetal cells or the amniotic fluid itself. For conditions like Tay–Sachs syndrome, they could identify the absence of the enzyme hexoseaminidase A in the amniotic cells.[18] This opened the floodgates for looking at lysosomal storage disease and other metabolic errors that produced harmful or lethal conditions in newborns. Those could now be detected by prenatal diagnosis. The last phase in diagnosis became available when DNA could be identified in fetal cells. This allowed all possible mutations, even when their role in metabolism or the tissue was unknown at a biochemical level, to be identified by prenatal diagnosis.

Accompanying the efforts to prevent babies with serious genetic disturbances were efforts at screening selected populations to identify carriers of ethnic recessive genetic disorders such as Tay–Sachs syndrome. Most American Jews who may be at risk for being a carrier chose to have such screening because the carrier incidence was about one in 25 persons of Ashkenazi ancestry. Those at risk used amniocentesis to abort fetuses with Tay–Sachs syndrome. Orthodox Jews chose the screening route and used arranged marriages, with the agent in charge of such arranged marriages selecting candidates who were not carriers if one candidate was a known carrier. The use of these genetic services for genetic mutations led to the virtual disappearance of live-born children with Tay–Sachs among Jewish couples, but it did not reduce the gene frequency of this mutant condition. Neither an aborted embryo nor a child dying at the age of 4 or 5 of Tay–Sachs syndrome will pass on the mutant gene to a new generation. Hence no change in gene frequency takes place.

In vitro fertilization is sometimes coupled with genetic screening. A four- or eight-celled embryo can have a blastomere removed and sampled for a genetic disorder, and because there are usually several fertilized eggs in any one IVF procedure, such an embryo with a genetic disorder will not be used for insertion into the uterus. This allows couples to put embryos that are either homozygous for the normal allele or heterozygous for the condition into the uterus.

Note that both prenatal diagnosis and IVF technology are widely known to the public in the industrial world. There is a public understanding of detrimental or harmful mutations that exist in the human population as well as knowledge that there are children born with extra chromosomes. Anyone attending a genetic counseling session will quickly realize that relatively modestly educated couples, most not having more than a high school education, will follow in a question-and-answer session with the counselor how the genetics works in their particular case. At the same time, old habits of thinking are hard to erase from public memory. The belief that mutations involve an unbroken parent-to-child transmission (i.e., mutant produces mutant) is part of popular culture and seen in the public fears of exposure to agents that cause mutations.

References and Notes

1. Betlach M. 1994. "Early cloning and recombinant DNA technology at Herbert W. Boyer's UCSF laboratory in the 1970s." Oral history interview by Sally Smith Hughes, University of California Oral History Project. Betlach was a technician at the time (she joined the laboratory in 1972) and later got a PhD in biochemistry. She describes the atmosphere of the laboratory, the roles of the participants, and her own contributions to the first plasmids containing horizontally transferred genes that constitute the "biological engineering" called "recombinant DNA technology." Also, see Bushman F. 2002. *Lateral DNA transfer.* Cold Spring Harbor Laboratory Press, Cold Spring Harbor, NY.
2. Carlson EA. 2006. *Times of triumph, times of doubt: Science and the battle for public trust.* Cold Spring Harbor Laboratory Press, Cold Spring Harbor, NY.
3. Watson JD, Tooze J. 1981. *The DNA story: A documentary history of gene cloning.* Freeman, San Francisco.
4. This required a teamwork of academic scientists like Boyer, who cofounded Genentech in 1976, and the financial backing to start up companies using recombinant DNA technology to do massive gene cloning and product production for medical use. Human growth hormone (somatostatin is a peptide

of 181 amino acids) was synthesized in 1977 and approved by the Food and Drug Administration in 1985. Human insulin contained two chains (21 and 30 amino acids) that were synthesized by recombinant DNA technology in 1978, also by Genentech.

5. The history of applied research reflects conflicting human interests. Some of the disputes involve patent infringements or fights over priority of invention of a product. Some involve regulation by agencies responsible for public health and safety. Others involve fears regarding unintended bad outcomes, as in the debates over recombinant DNA technology in the 1970s; alleged bad outcomes from users of products because of their response to the products (often a very small percentage of users have such reactions); fears of abuse of products for germ warfare or for cheating by athletes; or the pricing of products that makes them unavailable to the poor.

6. Saiki RK, Scharf S, Faloona F, Mullis KB, Horn GT, Erlich HA, Arnheim N. 1985. Enzymatic amplification of β-globin genomic sequences and restriction site analysis for diagnosis of sickle cell anemia. *Science* **230:** 1350–1354.

7. Carlson EA. 2001. Genetically altered organisms. *Dissent* (Winter 2001), pp. 55–61. Also, for an overview of the field and the controversy over genetically modified foods, see Pringle P. 2005. *Food, Inc.: Mendel to Monsanto—The promises and perils of the biotechnology harvest.* Simon and Schuster, New York. Another valuable book is Fedoroff N. 2004. *Mendel in the kitchen: A scientist's view of genetically modified foods.* National Academy Press, Washington, DC.

8. Capecchi MR. 1989. The new mouse genetics: Altering the genome by gene targeting. *Trends Genet* **5:** 70–76.

9. Cowan RS. 2008. *Heredity and hope: The case for genetic screening.* Harvard University Press, Cambridge, MA.

10. For a discussion of this controversy between Ira Kaplan and H.J. Muller, see Chapter 24, "Human mutations and the radiation danger," in Carlson EA. 1981. *Genes, radiation, and society: The life and work of H.J. Muller.* Cornell University Press, Ithaca, NY. Also see Kaplan's review of 30 years of work using this method: Kaplan II. 1957. The treatment of female sterility with X-rays to the ovaries and the pituitary with special reference to congenital abnormalities of the offspring. *Can Med Assoc J* **76:** 43–46.

11. Muller mentioned this in his course on radiation biology. B.P. Sonnenblick is known for his efforts in the state of New Jersey to pass radiation protection regulation laws. See Sonnenblick BP. 1955. On some aspects of the problem of human radiation protection. *J Newark Beth Isr Hosp* **6:** 31–43.

12. Reijo R, Lee TY, Salo P, Alagappan R, Brown LG, Rosenberg M, Rozen S, Jaffe T, Straus D, Hovatta O, et al. 1995. Diverse spermatogenic defects in humans caused by Y chromosome deletions encompassing a novel RNA-binding protein gene. *Nat Genet* **10:** 383–393.

13. David C. Page and his colleagues at MIT performed most of the analysis of the human Y chromosome, including note 12. Page showed that the Y was not an empty chromosome with just a sex-determining gene (*SRY*), but a uniquely

organized chromosome that had evolved with the X chromosome from a common set of autosomes some 300 million years ago. Inversions led to genetic isolation. The Y chromosome has 78 genes that produce proteins. Its structure includes eight large palindromic sequences that generate new mutations.

14. Steptoe PC, Edwards RG. 1978. Birth after the reimplantation of a human embryo. *Lancet* 2: 366. doi: 10.1016/S0140-6736(78)92957-4.
15. The options were provided by genetic counselors. Sheldon Reed at the University of Minnesota named the field "genetic counseling" in 1949. He wrote the first book on this new field: Reed SC. 1955. *Counseling in medical genetics.* Saunders, Philadelphia.
16. Lejeune J, Turpin R, Gautier M. 1959. Le mongolisme. Premier exemple d'aberration autosomique humaine. *Ann Génétique* **1:** 41–49.
17. Caspersson T, Zech L, Johnson C. 1970. Differential binding of alkylating fluorochromes in human chromosomes. *Exp Cell Res* **60:** 315–319. Crossen PE. 1972. Giemsa banding patterns of human chromosomes. *Clin Genet* **3:** 169–179.
18. The Tay–Sachs prenatal diagnosis was independently published in 1971. Navon R, Padeh B. 1971. Prenatal diagnosis of Tay–Sachs genotypes. *Br Med J* **4:** 17–70. O'Brien JS, Okada S, Fillerup DL, Veath ML, Adornato B, Brenner PH, Leroy JG. 1971. Tay–Sachs disease: Prenatal diagnosis. *Science* **172:** 61–64.

10

Mutation in Relation to Society

HAD YOU BEEN ALIVE IN THE MID-19TH CENTURY, a contemporary of Darwin in the life sciences, your views on mutation in relation to society would include some unease regarding the problem of what was then called degeneracy. Degeneracy was considered a declining departure from normalcy in physical appearance, health, or mental abilities and attitudes. There were books on degeneracy arguing that environmental agents were associated with the degeneracy of hatters using mercury, chimney sweeps dying of cancer (including an otherwise rare scrotal cancer), and slums creating an unwholesome population that was disproportionately alcoholic, feebleminded, syphilitic, tubercular, recidivist criminal, and prone to gambling, begging, prostitution, and other vices.[1] You would have probably read Charles Dickens's novel *Oliver Twist,* which addressed this problem of degeneracy, and, if you were sympathetic toward your religious upbringing, you would have considered Oliver's salvation from Fagin's criminal influence an act of divine grace.[2]

That there was degeneracy was not questioned. The debate was over its causation. Those with a Lamarckian view of heredity would have attributed it to a bad environment. The remedy was equally Lamarckian—provide a good environment. This was the stress of Richard Dugdale (1841–1883) in his 1870s presentation of the family he called "the Jukes," but his hope for a preschool program (importing the German model of *kindergarten*) was ignored in the 1880s, as were his recommendations of decent housing, a sound liberal arts education, and decent wages for the Jukes and their growing kindred.[3]

Not all Lamarckian biologists believed that the deterioration of a species was reversible. Benedict Morel (1809–1873) in the 1850s argued that pathological departures from the normal had reduced chances of transmitting

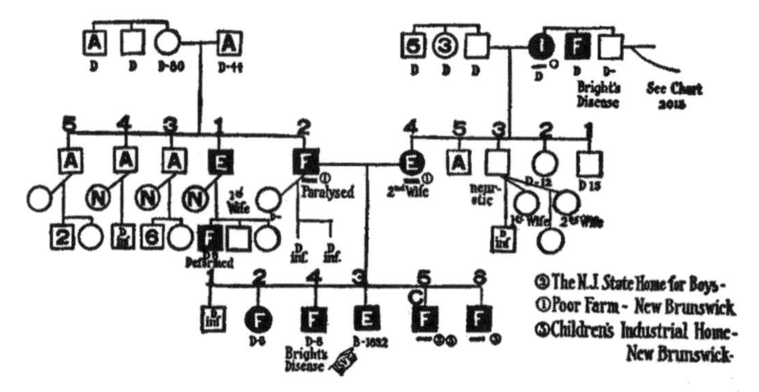

Fig. 38.—The product of a feeble-minded man (who has an epileptic brother) and his epileptic wife (whose father was insane and uncle feeble-minded); the first child died in infancy, the next two were feeble-minded and died young, the next is an epileptic at the New Jersey State Village; the next is feeble-minded, has a criminal record and is in the State Home for Boys; the last is feeble-minded and is in the Children's Industrial Home. Six in this family have been or are wards of the State. *A*, alcoholic; *C*, criminalistic; *D*, deaf; *E*, epileptic; *F*, feeble-minded; *I*, insane; *N*, normal. *SV* in the ☞ means an inmate of a State Village for Epileptics.

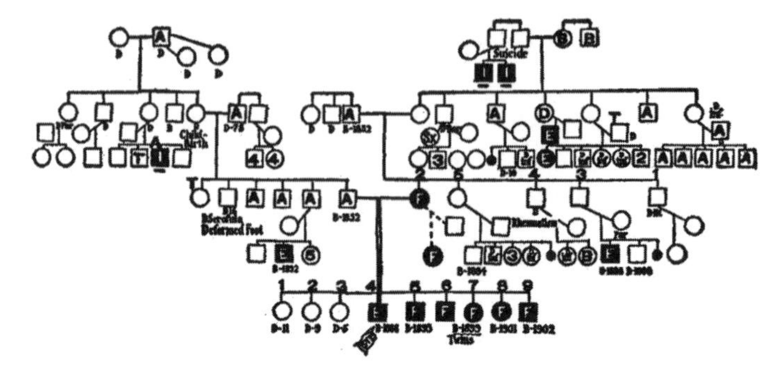

Fig. 39.—The central mating is that of a feeble-minded woman of an intensely neuropathic strain and an alcoholic man, who has 3 alcoholic brothers, father and grandfather alcoholic, an insane cousin and an epileptic nephew. The husband, though recorded as alcoholic, is probably also feeble-minded, at least all (6) of his children who survived were feeble-minded or epileptic. This chart shows 4 wards of the State and many others who should have been segregated. *A*, alcoholic; *B*, blind; *B*, (below), born; *D*, deaf; *D*, (below), died; *E*, epileptic; *F*, feeble-minded; *Ht*, heart-disease; *I*, insane; *Par*, paralysis, *Sx*, sex-offense; *T*, tubercular.

Problematic pedigrees foster the negative eugenics movement. In Charles Davenport's mind, the legitimacy of behavioral traits as an outcome of fixed Mendelian units was as clear to him as those he described for poultry and other animals he bred. The social traits of these kindreds tended to show dominant-like inheritance from both sides of the kindred, a situation that is also characteristic of environmentally acquired traits. Most dominant medical traits (such as Huntington disease or retinoblastoma) tend to show only one side of the family having transmitted the trait. (Reproduced from Davenport CB. 1911. Heredity in relation to eugenics. *Henry Holt and Company, New York; facsimile reprinted in* Davenport's dream: 21st century reflections on heredity and eugenics *[eds. Witkowski JA, Inglis JJ. Cold Spring Harbor Laboratory Press, Cold Spring Harbor, NY.)*

defects and they tended to die out, with few of these degenerate traits persisting more than five generations in a family. Other biologists chose a non-Lamarckian approach to degeneracy, especially after the publication of August Weismann's books on heredity in the 1880s and 1890s. Weismann argued that hereditary traits were independent of social environments. They were innate, and he assigned them to some fundamental attribute in the reproductive tissue, which he called the "germ plasm." The body tissues, known as the "soma" or "somatoplasm," could not pass acquired traits to the germ plasm. This was a sharp break between Weismann's idea of hereditary units and Darwin's model, where "gemmules" of the cells in body tissues were circulated and arrived at the germinal tissue and became assimilated. Darwin's model of "pangenesis," as he called it, was a mechanistic model for Lamarckian heredity and overcame the idea of "will" or "need" as the basis for Lamarckian traits. Pangenesis was dead by the 1870s because Francis Galton put it to experimental test using rabbits of different color patterns, and he showed no transmission of gemmules through blood transfusions. Darwin used a lame argument that he did not say that blood was the means by which gemmules were transmitted, but that was unconvincing to his readers.[4]

At the turn of the century in 1900, most biologists had been won over to Weismann's theories of a separate germinal heredity that was unresponsive to somatic environmental changes. The rediscovery of Mendelism provided a mechanism for transmitting hereditary units. This made it seem logical that if social degenerate traits were hereditary, they would be found in the germinal tissue and transmitted as Mendelian traits. Curiously, few biologists questioned the hereditary nature of degenerate traits in society. It was almost universally assumed to be true, most likely because of the multigenerational nature of pauperism, criminality, alcoholism, and other failings of society. For that reason, a movement begun in the early 1880s by Francis Galton, which he called "eugenics," found favor in society. Human betterment could be achieved through eugenic practices. These unwholesome traits quickly were assigned to reproductive isolation by retention of persons in asylums or prisons if they were psychotic, feebleminded, or criminal. For the far more numerous failings designated as pauperism, marriage laws or eugenic education concerning good marriages was sought. Where these failed to work, reproductive isolation was achieved by passage of compulsory sterilization laws. The first such law in the world was passed in Indiana in 1909 and signed by the Governor. These compulsory sterilization laws had the advantage of allowing degenerate persons back into society, but it kept them from reproducing their own kind. Hence the eugenic objective of the law would be achieved. As immigration from southern and eastern Europe to the United States swelled in the early 20th century, advocates of eugenics sought to isolate the U.S. stock from the "refuse of Europe" by restrictive immigration laws based on ethnic bias, and those were achieved during the 1920s.

Some worried about racial purity or preservation of the races and feared miscegenation. For well over a century, it was widely believed by European anthropologists that white people (called Caucasians or Aryans) were superior to other races and that racial physical differences reflected racial mental abilities and personalities.[5] In the United States that "scientific racism" found its way into state laws that restricted marriages to those of the same race. The fear was based on a false premise that hybrid or mixed heredities were inferior to that of either purebred stock. The biological model for this was the sterility or weakness of hybrids between species. But biologists since 1908 had shown "hybrid vigor" in maize and other agricultural commodities. The public, through their legislators, rarely on sound ground in using science for public policy, ignored that evidence, and few biologists used that argument to criticize efforts to ban racial miscegenation.

Genetics Is Purged of Eugenics

Eugenics was popular throughout the world in the early 20th century, and three international Congresses of Eugenics were held, one in London and two in the United States, the last in 1932. Why was it popular? Those who favored what was called "positive eugenics" believed that humans could control their own heredity and this could lead to future generations free of hereditary diseases and sharing higher intelligence, good health, and more talents than previous generations. This was a view that cut across ideological lines and was shared by communist, socialist, and capitalist scientists.[6] Those who favored what was known as "negative eugenics" had no such strong desire to reshape humanity but felt that their role was to conserve the human stock and its diversity by preventing its corruption through miscegenation or the breeding of the fit with the unfit. The negative eugenics movement in general appealed to capitalist societies and to scientists with a more conservative political outlook, especially throughout the 20th century.[7]

The negative eugenics movement was exemplified by the writings of Charles Davenport (1866–1944) and Harry Laughlin (1880–1943) at the

Eugenics Record Office located at Cold Spring Harbor in New York. Laughlin was the Superintendent. Many of his publications and statements to Congress reflected his biases, especially on class status, ethnic origins, and race. Critics of this eugenics movement thought of it as a "blame the victim" attitude. The paupers were held to be responsible for their poverty. The sick were accused of neglecting their personal hygiene and bringing on their own illnesses. The semiliterate were seen as lazy people who did not take advantage of opportunities like free public libraries.

Eugenics failed for numerous reasons. First, it was filled with abuses based on bias rather than convincing genetic evidence of impairment. Second, scientists realized that a simple model of "defective germ plasm" or a simple single-gene model of feeblemindedness, pauperism, criminality, or other vices was just not evident in the pedigrees that accumulated. Scientists also realized that genotypes had a range of expression depending on environments, and geneticists had long learned to use incubators or greenhouses to grow their experimental organisms under controlled conditions. Geneticists also showed that many traits had multiple-factor inheritance or were homozygous recessive, and sterilization of "the unfit" would take a very long time (several centuries) for social change to be noted. Third, the Great Depression made many social scientists aware of how vulnerable society was to calamitous economic shifts in fortune. When sizable portions of the population slide into unemployment because of economic hard times, it is hard to convince society that those unemployed should blame their genes. Fourth, the abuses by Germany during the Third Reich soured almost all geneticists and the public regarding eugenics after the war. Genocide, especially the Holocaust against Jews as a Nazi policy, was seen as abhorrent, and this resulted in international efforts to cite genocide as a war crime. Fifth, the fields of human genetics and medical genetics, purged of eugenics, established their legitimacy as sciences and shunned eugenic applications or ideology in their genetic services.

Mutation Is Perceived Differently in the Post–World War II Generation

Whereas eugenics virtually disappeared from science and public discourse after the end of World War II, the term mutation took on a different significance after the war ended. A new age of atomic bombs, nuclear reactors, and multiple uses of radiation in medicine, industry, and agriculture was ushered in. Concern for the genetic effects of radiation was initiated by

Hermann Joseph Muller in 1927 after he induced mutations with X rays and realized that the overwhelming number produced in fruit flies would be harmful to their survival; these mutations included recessive lethals (interpreted as mutations affecting the early stages of the fly life cycle, resulting in aborted embryos or larvae that failed to pupate). Also included was an abundant class of mutations, most of them recessive, that led to sterility; these were interpreted as mutations affecting the processes of egg and sperm production and the fertilization process. Among the visible mutations induced by X rays, Muller noted they were usually losses or deformities that would be considered pathological if they had occurred in humans instead of flies because of the loss of function of major organ systems. By the 1930s, Muller and colleagues were noting that X rays also produced an abundance of chromosome breakage, leading to structural rearrangements of the chromosomes, including inversions, duplications, deletions, and translocations. By the early 1940s, Muller and his students also identified abnormally repaired chromosome breakage that resulted in aneucentric chromosomes, especially acentric fragments and dicentric chromosomes, leading to breakage–fusion–bridge cycles, as Barbara McClintock called them. These induced breaks led to a class of mutation formerly called dominant lethals, but which were actually not changes in the individual gene but aneucentric rearrangements of irradiated gametes that aborted the early embryo.[8]

Muller was the first to recognize that radiation sickness could be interpreted as a consequence of aneucentric chromosome formation in dividing cells, making blood, skin, the intestinal lining, and reproductive tissue especially vulnerable to radiation exposure among those who survived the blast effects of the atomic bombs. These genetic effects of radiation also raised the possibility of protecting the public from unwanted radiation whether from wartime uses or peaceful uses of radiation in industry and medicine. But the genetic effects of radiation also led to controversy, with Cold War politics frequently trumping rational scientific discussion, leading to division between those who used the same data from worldwide fallout and came to opposite conclusions. Linus Pauling used the total estimated induced mutations worldwide to petition against further testing of nuclear weapons in the atmosphere. Edward Teller (1908–2003) calculated the number of mutations induced against the totality of persons born and showed that the individual risk was minuscule. Muller took an intermediate position. He recognized that the individual risk was low to a child born in the atomic age, but it was avoidable if both sides in the arms race would choose international control and sign a ban on atmospheric weapons testing. That position eventually won out, and the public concern over radiation damage began to recede.[9]

The Lysenko Controversy Involves a State-Mandated View of Heredity

But the Cold War intensified a growing debate among scientists in the USSR regarding variations in plants and other life. It began in the mid-1930s when Trofim D. Lysenko (1898–1976), an agronomist in Odessa, argued that he had transformed a strain of wheat by treating seeds with temperature shocks (germinating in a bed of ice). He claimed that this process, called vernalization, could change winter wheat into spring wheat, and this would enable more extensive planting of wheat in the USSR grain belts, especially in Ukraine. Lysenko continued his campaign to promote his techniques, which he called "Michurinism," enabling environmental methods to "shatter" the heredity of plants and to "educate" the altered plants with a new environment to the benefit of the collective farmers.[10] Opposed to Lysenko and his supporters were Nikolai Vavilov (1887–1943) and his supporters, including Muller, who was a guest investigator in the USSR in the 1930s. Lysenko lobbied the Politburo to withdraw the USSR from the International Congress of Genetics, and he began a campaign to change the teaching of genetics in schools and colleges. Michurinism, he claimed, was progressive and Darwinian, but formal genetics ("Weismannism–Mendelism–Morganism") was bourgeois, idealistic, or fascistic. Lysenko claimed that all heredity was malleable, including human heredity. Mutations were not fixed in the germ plasm but subject to environmental modification in the direction desired by the experimenter. It was not selection but some type of Lamarckism that Lysenko offered as his explanation for the environmental modifications. When Lysenko got the Communist Party to back his movement in a showdown meeting in 1948, a Cold War erupted among scientists, causing many left-wing geneticists to reconsider their sympathies for the USSR; but this also led to short-lived movements for Lysenko in France, Great Britain, and the United States. Eastern bloc countries in general favored teaching Lysenko's version of genetics, and Western bloc countries in general denounced Lysenkoism, as they liked to call the movement, as an intrusion by the state in science that was short-sighted, wrong, and dangerous to science.

By 1960, it was clear that Lysenkoism did not work, and despite efforts by Lysenko and his supporters to repress formal genetics, a new administration in the USSR replaced Nikita Khrushchev (1894–1971), Lysenko's last supporter, and Lysenkoism began to disappear from science in the Eastern bloc and the USSR. Even without this political shift, Soviet scientists who approached genetics from a molecular approach could successfully avoid Lysenkoism in the USSR. As long as they discussed their findings in the language of biochemistry and not in the classical terms of genes, point mutations, and the chromosome theory of heredity, their work could be published. Sadly, however, Lysenkoism severely damaged the life sciences in the USSR, which lost a generation of young scientists who avoided genetics as a field. Western geneticists who could not duplicate Lysenko's results suspected that the work was fraudulent, and later assessments by Russian scientists using documents of the time have confirmed that fraud or self-deception marked Lysenko's claims of bumper crop yields from vernalization.[11]

Mutation Is Applied to Human Genetics and Medicine

When Muller was elected to serve as President of the new American Society of Human Genetics in 1948, he gave an address to the members and urged them to establish a journal for their work that was purged of eugenics. He sought new research to establish a human cytogenetics, studies of populations to assess mutation rates, mathematical modeling of the distribution of genes in populations, studies of individual disorders to understand the methods of transmission of them, efforts to assess radiation damage in the populations of Hiroshima and Nagasaki, and other approaches that would help amass data and findings to shape the new field.[12] He told them to avoid applying that new knowledge to social problems because in 1948 too little was actually known regarding human genetics. That advice has prevailed, and human genetics became a science dominated in its first 20 years by those with PhDs. Few physicians were among those founding geneticists. But that changed by the 1970s as numerous medical applications became possible.

Public interest had shifted from concern regarding nuclear war and its worries about induced mutations to concern regarding their present-day realities. Human genetics had spawned the field of genetic counseling. This was guided by the eugenic purge, and its practitioners were urged to provide information and to respect the autonomy of the clients seeking that information. Respecting their reproductive decisions regarding how to use the information was part of an increasing tendency in American and

European society to respect the individual as an autonomous decision-maker, and not treat him or her as a subject to be guided by the state or by a profession, especially in matters involving privacy such as reproduction and medical health.

Medical genetics was also spawned from human genetics as the discovery of human chromosome disorders proliferated in the late 1950s and early 1960s. This led to techniques of amniocentesis or detection of chromosomal abnormalities in the amniotic fluid obtained from pregnant women in their first trimester of pregnancy. Amniocentesis is an invasive technique, and thus medical training was required to perform it successfully. As MDs took an interest in human genetics, they began looking at amniocentesis as a tool for the study of biochemical disorders. Very quickly, a new field of genetic screening emerged to detect carriers in the population who were at risk for having children with lysosomal storage disorders, metabolic deficiencies, and other conditions that would show up in the amniotic fluid or the cells of the fetus. Although trisomy 21 or Down syndrome is not a mutation in the sense of a change in the individual gene, it was perceived by many as a type of mutation or variation that they wished to avoid bringing into the world. Point mutations such as those associated with Tay–Sachs syndrome, cystic fibrosis, retinoblastoma, sickle cell anemia, or Huntington disease rapidly became known to the public as preventable through prenatal diagnosis.

Unlike the history of mutation seen from a social vantage in prior generations since Darwin, mutation since the advent of genetic counseling and genetic services such as prenatal diagnosis, genetic screening, and in vitro fertilization has been strongly influenced by religious opposition to its primary uses—the prevention of the birth of children with genetic disorders (all medically significant gene mutations, chromosomal rearrangements, or ploidy abnormalities). That opposition is based on several religious values. Paramount is the issue of elective abortion. To many religions, this is equated to murder, and fertilized eggs are given the status of personhood. A second issue (more typically associated with the theological traditions of the Catholic Church) is natural law, or how the process is carried out. Most medical technologies associated with infertility attempt to bypass the blocked normal path through sexual intercourse and substitute sperm donation, fertilization in a dish, masturbation to obtain sperm, suction to obtain eggs, and other methods that can bring about fertilization in laboratory conditions. Even if no death of zygotes occurs, such practices were forbidden by Catholic doctrine. Intentionality is also raised as an objection, but less frequently in the 21st century than in the last half of the 20th century. In the mid-20th century, it was not uncommon to see arguments by Church authorities that use of any means (other than abstinence) of birth control was morally unacceptable if the intention was to prevent conception, and even when no artificial means was used, there were limits imposed on the duration of abstinence in a marital state during the reproductive life of the couple. The use of modern technology in human reproduction also varies with the religious practices of other populations around the world, some seeing no moral difficulties with any of the new technologies and others shunning them.

It is not just religious beliefs that make it difficult for some people to use medical technologies for their infertility or genetic risk concerns. Some object to its widespread use on philosophical grounds and see the encroaching medicalization of reproduction as leading to "eugenics through a back door" or discrimination against the handicapped or a warped value system that favors "designer children" and the imposition of self-centered values on future generations. Some even argue that children with birth defects or expressing genetic syndromes are not disabled, handicapped, retarded, or impaired as persons. They believe that these are discriminatory terms reflecting the biases of those who need to be educated. Instead of looking on these children as abnormal, they see them as "differently abled" and believe that society should shun prevention of these births and promote compensatory education and other opportunities for these children to integrate into society. Others feel sympathy to the prevention of future births by genetic screening and prenatal diagnosis with elective abortion; but, for those already born, they desire the very measures their opponents advocate. Most people prefer options to choose rather than government mandates that limit such choices.[13]

Mutation Shifts to Safety Concerns in an Age of Complex Technologies

If the public perception of mutation in the 1850s was largely focused on degeneracy theory, in the early 1900s on eugenics, after World War II on Cold War concerns and the nuclear age, and in the last half of the 20th century on new reproductive technologies and the prevention of infertility and genetic disorders, what are the concerns about mutation in the molecular age that will dominate the 21st century? We can see the present tensions in the fights over genetically modified foods. A segment of society favors the natural over the artificial. To most scientists, this is a false distinction because most of what is considered natural today (e.g., public health, immunization against infectious diseases, family planning, universal literacy, health insurance, Social Security pensions for the retired, hybrid corn, genetically selected strains of food crops and animals, or breeding dogs and other pets) was won with difficulty over many years, and only after one or more generations grew up with these services and activities were they taken as natural. A similar outcome is likely to happen with genetically modified foods, as the early mistakes and omissions in its launching are not repeated and the successes of the products and their safety sink in to those using these new practices. The popularizing of nucleotide sequence analysis in television crime shows is making the public aware of mutation as an alteration in the individual gene and that a string of just a few genes will be sufficient to distinguish one person from another. The initial fears of runaway epidemics from genetically recombinant microbes used for commercial production of products like insulin or human growth hormone have largely disappeared.

The public still has shallow interpretations of mutation, as seen in the origin of superhero comic strip heroes from exposure to mutagens (X rays, toxins). But it still lacks the insight into the largely heterozygous state of most significant induced mutations in individuals. It still tilts toward a like-for-like mode of inheritance that was almost universally accepted a little more than 100 years ago. Too many people think of induced mutations as having equal probabilities of being beneficial or harmful. Too many people have the same fears of low doses of radiation as they do of high doses of radiation. Most people are unaware of what categories of chemicals are likely to be mutagenic. Fortunately, a half-century of bombardment by public health articles has convinced most of the literate world that smoking tobacco results in mutations in respiratory tissue that lead to cancers. The cavalier response of the past ("you have to die of something") is less frequently heard today. We may not be able to predict how the public sees mutation in the coming generations, but the trend since the 1850s has shifted toward a more solid science based on experimentation and less on philosophical, ideological, or socially biased beliefs.

References and Notes

1. For a history of degeneracy theory and its relation to the eugenics movement, see Carlson EA. 2001. *The unfit: A history of a bad idea.* Cold Spring Harbor Laboratory Press, Cold Spring Harbor, NY.

2. But the pupils of Fagin in his school for future criminals were seen as malleable and could be rescued before they became permanently corrupted like Fagin and Sykes, the villains of the novel. For Dickens, societal reform, not reliance on the gallows, was the answer to the vices that abounded among the poor.

3. Dugdale R. 1877. *The Jukes: A study in crime, pauperism, disease, and heredity.* Putnam's Sons, New York.
4. Galton F. 1871. Experiments in pangenesis, by breeding from rabbits of a pure variety, into whose circulation blood from other varieties had previously been largely transfused. *Proc R Soc Lond B Biol Sci* **19:** 393–404.
5. Gould S. 1981. *The mismeasure of man.* Norton, New York.

6. Among the "social progressives" favoring positive eugenics were H.J. Muller, A.S. Serebrovsky, Julian Huxley, and J.B.S. Haldane.
7. Among the supporters of negative eugenics (eliminating the unfit) were R.A. Fisher, Leonard Darwin, Erwin Baur, Charles Davenport, David Starr Jordan, and Harry Laughlin. The negative eugenics movement far outnumbered the positive eugenics movement in adherents. Both movements were compromised and sullied by Nazi German policies that focused on race (Aryan superiority) and brutal treatment of those deemed inferior (especially anti-Semitism culminating in genocide).

8. McClintock B. 1938. *The fusion of broken ends of sister half-chromatids following chromatid breakage at meiotic anaphases.* University of Missouri, College of Agriculture, Research Bulletin 290, pp. 1–48. University of Missouri, Columbia, MO. Muller HJ, Pontecorvo G. 1942. The surprisingly high frequency of spontaneous and induced breakage and its expression through germinal lethals. *Genetics* **27:** 157–158.
9. Muller HJ. 1950. Radiation damage to the genetic material. *Am Sci* **38:** 35–39; 399–425.
10. Roll-Hansen N. 2005. *The Lysenko effect: The politics of science.* Humanity Books, Amherst, NY.
11. Medvedev R. 1969. *The rise and fall of T.D. Lysenko.* Columbia University Press, New York.
12. Muller HJ. 1949. Progress and prospects in human genetics. *Am J Hum Genet* **1:** 1–18.
13. Mansfield C, Hopfer S, Marteau TM. 1999. Termination rates after prenatal diagnosis of Down syndrome, spina bifida, anencephaly, and Turner and Klinefelter syndromes: A systematic literature review. *Prenat Diagn* **19:** 808–812. In their study, Mansfield et al. state that 92% of women with a prenatal diagnosis of Down syndrome elected termination by abortion. The lowest incidence for termination of the conditions they studied was 58% for pregnancies diagnosed with Klinefelter syndrome (47,XXY).

11

Mutation in Relation to History and Philosophy of Science

MY INTEREST IN THE CHANGING CONCEPT of mutation arose from several sources. First was Hermann Joseph Muller's course in Mutation and the Gene that I took in the fall semester of 1954. I was stunned by the number of contending ideas and terms that Muller introduced as he described the changes leading to Edmund Beecher Wilson's and Thomas Hunt Morgan's thinking, as well as that of their earlier formative generation.[1] Muller also followed up the numerous struggles with his contemporaries as they tried to work out the interpretations of genetic phenomena from a then meager amount of facts and experiments. It was this course that shaped my scientific philosophy that, at least for genetics, revolutions do not occur through "Kuhnian" paradigm shifts (then called scientific revolutions), but rather they occur incrementally from the findings of numerous experiments, new techniques used to study heredity, and new interpretations put forward to stimulate more experiments.[2] Of course, the term paradigm shift did not exist in 1954, nor did Muller have a respect for the philosophy of science, a prejudice he had acquired from his teacher, Morgan. Both Morgan and Muller considered themselves to be reductionists, although Muller thought Morgan's version was a cruder mechanism lacking an appreciation for complexity. Both would have agreed that experimentation was necessary to test ideas, and both would have agreed that all complex living phenomena could be resolved by suitable experiments with suitable technology. They would both have rejected the idea that "the organism is greater than the sum of its parts" as holistic and inviting of mysticism. What they believed was that the organization of parts was as essential as the isolation of the parts in a complex system, but that the organization was itself subject to reductionist approaches. They differed in the role of hypotheses. Muller thought

9-1-48

Is the gene increased by

instead of

Would have to postulate an equilibrium volume to keep the gene stable - Muller does not believe in this

What would prevent mutation occurring in these individual molecules.

1. wouldn't they all eventually become different.
2. however some postulate a special situation in which a single mutation effects them all.
3. This is one of the reasons why Muller wanted to do X-ray work.

Genomere Hypothesis of Anderson & Whaley

1. many parts to a gene - one part can change and other not.
2. hypothesis set up to explain unstable genes in corn.
3. segregation in daughter chromatids - random distribution of particles
4. patches would be sorting out of mutant particles - not new mutations
5. Demerec - unstable gene in Drosophila virilis - believed for a while in genomere hypothesis
6. X-ray work was to see if you can produce exact mutations in genomeres. Should result in mosaics (irregular distribution). However X-rays gave no such results.

Results in a choice of the 3 following explanations

1. All genomeres affected in a similar fashion
2. Have only 1 particle
3. More than one element but some regular method of distribution

Half and Half Mosaics were found. Could be explained on basis of

1. 2 genes in sperm. already reduplicated
2. may be after effects - takes place after fertilization

1/4 & 3/4 not found.

James Watson's notes from Muller's course on Mutation and the Gene in 1948. This page from his class notebook shows Muller's motivation for seeking mutations with X rays. First was a need to clarify the process as specific to a gene and second was to rule out "genomeres" or variable patches of mutant and nonmutant particles alleged to be the source of spotting and streaking in corn kernels. Muller's work in 1927 showed the mutations were specific to the gene and such mosaicism as did occur was limited to a single generation. (Reprinted from Watson's class notebook for Muller's course in Mutation and the Gene [1948]. Courtesy of the Watson collection, Cold Spring Harbor Archives.)

that they were essential and if correct deserved as much recognition as the performing of an experiment that verified them. Morgan was more skeptical of ideas and interpretations, considering them too numerous and easy to toss out to merit credit for discovery.[3]

My interest in the history of science arose from my love of reading books about the history of ideas and fields such as medicine, law, or evolution, a habit I acquired in high school. When I was an undergraduate at New York University, my history teacher was Wallace K. Ferguson (1902–1983), a historiographer and scholar of Renaissance studies (he preferred to call it the transition between medieval and modern society).[4] He invited me to attend a meeting of the History of Science society with George Sarton (1884–1956) presiding. Most of the field then was limited to science before the 19th century. When I entered graduate school, I wanted to work with Muller, and had written to him before making a commitment to Indiana University. Although I put my interest in the history of science aside as I became a geneticist, I never lost my interest in it. To a great degree, I and other students of Muller adopted his reductionist outlook on science, but I was more moved by the historical approach that Muller used than many of the graduate students. I did not pursue it until 1965, when I had a sabbatical leave at Woods Hole, Massachusetts, and wrote *The Gene: A Critical History*.[5]

In 1972, Bentley Glass (1906–2005), then the Vice President of Stony Brook University, asked me to look into the formation of a history and philosophy of science program for the University. I traveled to meet several of the leaders in that field, including Thomas J. Kuhn (1922–1996), who was then at Princeton University. During one lunch meeting, I asked him about the relative scarcity of paradigm shifts in the life sciences and why the physical sciences differed in that way from the life sciences. He felt that this was largely because so much of life science is descriptive compared to the physical sciences, which have had a long tradition of theoretical science. He also said that his interest was now more focused on the way terms undergo changes and that the vocabulary of science merits attention. That thought resonated well with my experience in Muller's courses, and this book is a product of that attempt to make sense of how we describe change in the living world.

I began the history of the idea of mutation with the decades of Darwin's youth because there was not much of a scientific interest in mutation before his introduction of the idea of fluctuating variations as the basis for natural selection. In antiquity, mutation was associated with the irony of gods at play with their human creations. Ovid's *Metamorphoses* describes changes that are associated with human behavior.[6] A weaver becomes a spider. The children of Hermes and Aphrodite become entwined as Hermaphroditus. There is more comedy than tragedy in Ovid's stories. But the Greeks also had humans undergo mutational changes through their personality flaws. Acts of hubris (a sort of defiance of the balance of the universe) led to multigenerational tragic consequences, as in the House of Atreus and its stories told by Aeschylus, Sophocles, and Euripides.[7] Even in medieval times, that sense of divine retribution was at play in the interpretation of human monstrosities, many of them associated with copulation with animals or the devil.[8] Others were actually thought to be signs or omens used to foretell tragic events for the royal family or the nation. None of this was science, but science was not fully independent from pseudoscience until the 16th century, when the works of Copernicus and Galileo had a powerful influence on the use of reason and a respect that the universe runs by laws, not the whims of deities.

The vocabulary of breeders dominated the education of Darwin in his formative years as a scholar in natural history. Variations, rogues, sports, breeds, atavisms, and monstrosities were the major ideas of how life differs. What caused these changes and in what fundamental ways they differed from one another were unknown. The cell theory does not enter until 1838 and does not become elevated to a doctrine until 1855. Chromosomes will not appear until 1878. Mitosis and meiosis will not be worked out until the 1880s and 1890s. Darwin's use of the terms "fluctuating variations" and "plasticity" described first the raw material on which natural selection acts and the assumed physiological basis (mostly domestication) that led to numerous variations among the breeds of finches, parakeets, poultry, pigeons, and other birds that Darwin studied in his visits to breeding exhibitions. He introduced the term "gemmules" for hypothetical hereditary units, and he felt disappointed that his provisional theory of pangenesis failed to meet its first experimental test. Darwin died without a solution to the origin of variations and how they are transmitted.

The idea of "physiological units," as Herbert Spencer called the inferred hereditary units, or "elements," as Mendel called them, may have been stimulated by John Dalton's atomic theory of how molecules are composed and the cell theory, which saw organisms as communities of smaller units, or cells. Reductionism was permeating the sciences in the 19th century, and holistic approaches were struggling to survive, especially in Germany, where holism was part of Johann Wolfgang von Goethe's (1749–1832) Romantic tradition, but the German universities were using reductionist approaches (especially experimental cell biology) to work out the components of the cell and their functions.[9]

After Darwin's death, there was a shift by a new generation of scholars to an interest by William Bateson in England and by Hugo de Vries in Holland to discontinuous variation. Instead of seeing sports as monstrosities, Bateson saw them as possible progenitors of body form and the evolution of higher categories than species. He added meristic and homeotic mutations to the vocabulary of heredity.[10] de Vries saw his variations in *Oenothera* as experimentally proven instances of new varieties and even new species arising by saltations or sudden departures from the norm. Gregor Mendel had the misfortune to be ignored. His statistical study of fixed ratios for discontinuous traits seemed unreal to his contemporaries in biology, who were used to diversity in kinds and individuals. They also doubted that biology had laws because of this diversity.

The collision between these heretical views and the orthodoxy of classical fluctuating variations among the biometricians led to a bitter dispute that dragged on for more than 20 years. Neither side won this debate because a new intruder from across the Atlantic entered the fray. Wilson and his students combined the newly discovered Mendelism through breeding analysis with cytology and came up with the chromosome theory of heredity. Wilson's ideas were absorbed by his colleague Morgan once he switched loyalties from de Vries's saltationist outlook to the new processes of sex-limited inheritance and crossing over that he found in fruit flies. Morgan expanded his research with a productive crop of students, especially Alfred Henry Sturtevant, Calvin B. Bridges, and Muller, who fleshed out classical genetics and produced a torrent of new terms—multiple alleles, crossover recombinants, nondisjunction, chief genes and modifiers, chromosomal rearrangements, reverse mutations, dosage compensation, position effect mutations, coincidence and interference in crossing over, X-linked lethal mutations, and pleiotropism. Those terms prevailed, and the older terms of the 19th century largely withered into oblivion.

A similar shift of terminology took place after 1953 with the introduction of the double-helix model of DNA and the numerous lines of evidence that hereditary material, the genes, were composed of DNA.[11] The terminology changed to base substitutions, transitions, transversions, frameshift mutations, operons, *hox* genes, epigenetic mutations, X-inactivation, thymine dimers, repair enzyme mutations, transposable elements, vectors, recombinant DNA technology, directed mutation, knockout mutations, SNPs, and comparative genomics, among many terms more descriptive and specific than their classical genetic counterparts or antecedents.

I prefer to think of these transitions in vocabulary as the accumulation of incremental knowledge based on new techniques and experiments producing new data that seeks explanation. Eventually they accumulate into a corpus of work like Darwinian variations of the 1850s to 1880s, the biometric statistical vocabulary of the 1880s to early 1900s, the Batesonian discontinuous variation vocabulary of the 1890s to 1920s, the classical genetic vocabulary of 1910 to 1950, and the molecular vocabulary of the 1950s to the 1990s. Since the beginning of the 21st century, the vocabulary to describe mutation has been shifting to a blend of computer-based terminology for gene sequencing and an advanced mathematical language to describe comparisons of gene sequences. These mathematical methods also allow descriptions of epigenetic activities in the life cycle and explore how protein domains can be combined to produce more proteins than genes available to account for them. The new vocabulary is driven by new technologies to study heredity at ever smaller levels—from cells, to chromosomes, to genes, to DNA molecules, to nucleotide sequences, and to the numerous components and interactions of those components of genes and their products. In following the story of mutation, step by step, we can see these transitions as old terms led to new ones. In the SNPs of the introns and exons of today's genes, there are still echoes of Darwin's fluctuating variations.

References and Notes

1. Jim Watson's notebook for the course when he took it in 1948 is in the Cold Spring Harbor Laboratory Library archives.
2. Kuhn TJ. 1962. *The structure of scientific revolutions.* The University of Chicago Press, Chicago.
3. My information on the style of Morgan's teaching and research are from an interview I had with Fernandus Payne at Indiana University about 1968. Payne studied with Morgan and Wilson about 1907–1910. Muller felt disadvantaged in Morgan's laboratory because both Sturtevant and Bridges received financial support from Morgan, but Muller did not. This feeling of injustice spilled over into priority for the ideas that entered the work of the fly lab. Later Muller felt that Bridges had learned of Muller's work on bar eyes from visitors to Muller's laboratory in Moscow, who then stopped at Caltech. The speed of Bridges's publication (a few days after submission) added to that suspicion. My own feeling is that Painter, Bridges, and Muller all thought of using Painter's techniques to look at bar as a possible duplication.

4. Ferguson WK. 1948. *Renaissance in historical thought: Five centuries of interpretation.* Houghton Mifflin, Boston.
5. Carlson EA. 1966. *The gene: A critical history.* Saunders, Philadelphia.
6. Ovid. 1955. *Metamorphoses* (transl. Humphries R). Indiana University Press, Bloomington, IN.
7. Mullahy P. 1948. *Oedipus: Myth and complex.* Hermitage Press, New York.
8. Paré A. 1982. *On monsters and marvels* [ca. 1580] (transl. Pallister JL). The University of Chicago Press, Chicago.
9. Nordenskiöld E. 1928. *History of biology.* Knopf, New York.
10. Bateson W. 1894. *Materials for the study of variation: Treated with special regard to discontinuity in the origin of species.* Macmillan, London.
11. Olby R. 1974. *The path to the double helix.* University of Washington Press, Seattle.

Glossary of Terms Associated with Mutation

Accessory chromosome: A chromosome found by Clarence Clung in male but not female cells of grasshoppers that may have served as a sex determinant.

Adaptive enzyme formation: A process in which all cells in a population exposed to a new metabolite will switch to producing the enzyme needed to digest the new metabolite.

Allele: Originally, **allelomorph**. One of two or more states of a hereditary unit, usually the normal (wild type) and one of its mutant forms.

Allelic contamination: An erroneous belief that the presence of two alleles in a heterozygote would lead to small bits of each undergoing an exchange, resulting in a spotted or variable expression of that gene.

Aneuploidy: Changes in expression of genes associated with an excess or deficit of one or more chromosomes. In animals it is usually incompatible with embryonic life, but in flowering plants it often is a means of new variety or species formation.

Aperiodic crystal: The gene as inferred by Erwin Schrödinger, retaining its crystalline capacity to replicate but having its particular genetic function transmitted.

Atavism: A reversion to an ancestral type by one or more offspring when two breeds or varieties are crossed to one another.

Autonomous mutation: A gene mutation that retains its expression when an organism has a mosaic distribution of cells containing it.

Balanced lethals: A status that permits a heterozygous dominant mutation (that is homozygous lethal) to breed true if a second lethal arises on a homologous chromosome very close in map distance to it, or if one of the chromosomes bears an inversion and recessive lethal, preventing recombination between the two homologs.

Balanced polymorphism: The likely status of genes if a heterozygous individual is more likely to survive than either of its homozygous alleles, increasing the number of heterozygotes in the population.

Base substitution: The molecular basis of a point mutation, in which a single nucleotide is replaced by another nucleotide in its DNA.

Bicolorism: The status of a gene's expression in relation to dose. In the case of eosin, an X-linked mutation, males (which have one X chromosome) are lighter in eye color than females (which have two X chromosomes).

Breed: A collection of traits in a population maintained by selective breeding.

Bud sport: A new variation arising from a bud of a plant that can be perpetuated by grafting.

Chemical mutagen: A molecule known to induce mutations.

Chief gene: A gene for a variable trait such as beaded or truncate wings, which is modified by modifying genes that increase (intensifiers) or decrease (diminishers) its expression.

Chromosome rearrangements: After breakage, chromosome fragments may be deleted, duplicated, inverted, or translocated. Some rearrangements lead to infertility, aborted embryos, or malformations in subsequent generations. They may also be a source of new varieties and play a role in evolutionary change.

Cistron: The gene as a functional unit defined by a *cis–trans* test of its alleles.

Codescript: The term used by Schrödinger for the inferred genetic code enabling a gene to carry out its function. In Schrödinger's model it was analogous to a Morse code.

Codon: A sequence of three nucleotides specifying a specific amino acid or a stop signal for translation.

Collinearity (also colinearity): The belief that the sequence of nucleotides in the DNA of a gene corresponds to the sequence of amino acids of a protein. This would involve three nucleotides per one amino acid, which holds true for bacteria and viruses but not for eukaryotic genes.

Comparative genomics: The use of computers to analyze sequence similarities and differences among two or more genomes that have been fully sequenced. The results can then be assembled into phylogenetic trees.

Complementation: In a pseudoallelic series, two alleles are said to be complementary if their *trans* alignment shows a normal phenotype. If m^1/m^3 is phenotypically normal but m^1/m^2 and m^2/m^3 are mutant in phenotype, then m^1 and m^3 show complementation.

Complete mutation: A newly arising mutation that is expressed in the soma and transmitted by the germline.

Complex locus: A pseudoallelic series in which allelism, pseudoallelism, and complementation are all demonstrable.

Congenital peculiarities: A variation that arises suddenly to produce a malformed organ.

Continuous variation: A trait that often is distributed as a Gaussian curve among the members of a species or a population of that species.

Convariant replication: The term used by Nikolay Timofeef-Ressovsky to represent H.J. Muller's concept that the fundamental property of the gene (and life) was its ability to copy and transmit its newly arising variations.

Coupling and repulsion: The tendency of two traits to stick together (coupling) or remain apart (repulsion) after their temporary association in a hybrid.

CRM mutation: A point mutation resulting in a gene product that may lose its antigenic specificity. CRM stands for cross-reacting material. CRM-positive mutations retain the antigenicity; CRM-negative mutations lose it.

Crossing over: The process by which genes on a pair of homologous chromosomes are shifted into recombined alignments.

Cryptic mosaicism: The inferred presence of mutant tissue in an apparently normal individual whose germinal tissue does not contain cells with the mutant condition.

Defective germ plasm: A belief (1880–1945) that degenerate traits in the germ plasm will be transmitted to the progeny.

Degeneracy theory: A belief that bad environments or habits lead to physical or mental impairment that was transmitted to the offspring.

Delayed effects: With chemical mutagens, it may take one or more generations for a mutation to be fixed as a complete mutation. Some delayed effects may result from a sorting out of mutant and nonmutant strands of DNA.

Directed mutation: A mutation whose sequence can be altered as desired and then inserted and tested for its function.

Discontinuous variation: A stable departure from a normal trait that can be transmitted with no apparent modification to one or more subsequent generations.

Domain: A region of a protein whose sequence is found in other proteins. Members with a shared domain often have common functions or cognate functions.

Dominant lethal mutation: An aborted embryo usually arising from a chromosome break leading to a dicentric chromosome in the zygote.

Dominant mutation: A trait that is fully expressed in the heterozygote.

Dosage compensation: The process equalizing the dose difference of X-linked mutations so that they show the same expression in males as they do in females.

Element: The term Gregor Mendel used to describe an inferred unit associated with a trait he studied.

Ethnic disorder: The excess frequency of a mutant gene found in an ethnic population that proliferated as a founder effect in some distant past or in which a specific type of selection allowed the mutant gene to proliferate.

Ever sporting displacement: A variable trait associated with a position effect, such as a chromosome rearrangement placing heterochromatin near a gene.

Fluctuating variations: Small or subtle changes in a trait among a population that serve as a basis for natural selection.

Founder effect: The presence of a rare mutation in excess as an initially small population having such an allele greatly increases in size in a new territory or through reproductive isolation.

Four-point test: In a multiple allelic series *m*, two alleles are marked with identifying genes and arranged so that am^1b^+/a^+m^2b will yield am^1m^2b or $a^+m^+b^+$ as recombinant pseudoallelic progeny, indicating that m^1 is to the left of m^2.

Frameshift mutation: A point mutation adding or subtracting one or two nucleotide pairs from a codon in a gene.

Gemmule: A unit of inheritance from a cell, proposed by Darwin, that circulates from cell to cell throughout the organism.

Genetically modified foods: The construction of plants or animals, using recombinant DNA technology, in which one or more genes are inserted that confer medical, pharmaceutical, or commercially valued traits. It may be described as genetic engineering of crops and animals that are used for food.

Genetic counseling: The preparation of a pedigree and the genetic testing of individuals who seek advice on their reproductive options. It is nondirective, and the information given to clients is theirs to use as their own values dictate.

Genetic fine structure: The saturated map of a gene in viruses or bacteria obtained by using pairs of alleles and appropriate markers to map them. It corresponds to the nucleotide sequence.

Genetic fingerprinting: A technique used to digest a protein into smaller fragments, separating them into visible spots on chromatic paper, and isolating and analyzing the individual spot to reveal its amino acid sequence. The specific lesion in HbS was identified by this means comparing HbA and HbS. In forensic studies it is used to identify paternity or crime scene presence in civil cases or for criminal cases.

Genetic isolating mechanisms: The production of large inversions or translocations in a population to prevent fertile offspring with a cognate population lacking these rearrangements.

Genetic load: The sum of deleterious mutations in a population that is measured as an equilibrium of incoming spontaneous mutations and outgoing genetic deaths of individuals who fail to leave progeny.

Genetic marker: A gene used to tag a chromosome region in which a nearby unrelated gene can be followed or identified.

Genetic screening: The testing of cells from skin, blood, or mucous membranes to reveal the presence of mutant recessive alleles in a person who does not express that disorder.

Genotype: The inferred genetic composition of an expressed trait requiring genetic analysis for its determination.

Germ plasm: A theory that hereditary units in reproductive cells are unaffected by changes taking place in the body cells.

Hardy–Weinberg law: A basic principle of genetics demonstrating that two alleles retain their percentage in subsequent generations if mating is random and on a large scale and if neither allele has a selective advantage.

Heredity: [19th century definition] The normal traits of a species and their transmission to their progeny. [20th century definition] The variations, normal and mutant, of a species and their transmission to their progeny.

Heterochromosomes: The name assigned by Nettie Stevens to chromosomes associated with sex in *Diptera* and other insects.

Heterosis: The occurrence of hybrid vigor in the progeny when two inbred strains are crossed to one another.

Heterozygous: The hybrid state in which a cell or organism bears a recessive allele and a dominant counterpart. Symbolically, Aa.

High blood: A breed whose desired traits are maintained by intense inbreeding.

Homeobox: A region of a gene associated with the control of segmentation in a developing embryo.

Homeosis: Designated later as homeotic mutants in which a developmental rudiment is located in an atypical place resulting in conditions like antennapedia in fruit flies, where a leg appears instead of an antenna on the head.

Homozygous: The status of a cell or organism in which both alleles are identified symbolically as AA or aa.

Intercalating agent: A chemical mutagen capable of producing frameshift mutations by inserting between base pairs and distorting the shape of the DNA molecule.

Intracellular pangenesis: A belief that the hereditary units reside in the cell, do not leave that cell, and are probably associated with the chromosomes of the cell.

Intragenic minute rearrangements: Mutations that may be mapped as point mutations but that, through salivary gland chromosome analysis or through DNA sequencing, turn out to be small inversions, deletions, or duplications.

Knockout mutation: An altered DNA or mutant variant of a gene is inserted into a stem cell (blastomere) and with appropriate markers can be followed to see if it is present in the organism that emerges. By use of the marker genes, the altered gene can be followed in a later generation in the homozygous condition and its phenotype and time of action determined.

Latency: The failure of a trait to express itself in one or more generations before it reappears in a later generation.

Left–right test: A method used by Muller to identify the boundaries of genes. Two different inversions involving breakage in the *y-ac-sc* region were allowed to recombine, and the crossover products failed to show a gain or loss of genetic material.

Like-for-like inheritance: A popular belief that inheritance is directly transmitted from parent to child.

Lysenkoism: The doctrine of Michurinism proposed by Trofim Lysenko advocating environmental means first to "shatter" heredity and then to "retrain" the plant or animal so that it would acquire beneficial traits that can be passed on.

Mendelian trait: A trait whose transmission shows a segregation ratio of 3 dominant to 1 recessive expression of the traits when two hybrids for the trait are crossed.

Meristic variation: Changes of number for structures or organs, including vertebrae and teeth, or multiplication of entire structures such as digits and limbs.

Milli-micro-molar reactions: An interpretation inferred from the functioning of pseudoalleles by Edward Lewis and Guido Pontecorvo, who each assumed that the products of such complex loci required rapid local interaction of the products of each component of a multiple allele system of tandem genes.

Molecular clock: Measuring the age of genes or DNA based on the number of nucleotide pair differences between species or higher taxonomic groups.

Molecular disease: A term used by Linus Pauling to show how an inherited disease could be followed from a defect in a gene, to a defect in its protein product, to a defect in the cell organelle in which the product plays a role, and so on up the level of tissue, organ, organism, and population.

Muller's doctrine: All genes arise from preexisting genes except for the first gene or molecule that could replicate its variations.

Muller's ratchet: If recombination is suppressed in an organism or in a pair of homologous chromosomes, a steady accumulation of detrimental mutations will occur. Muller's ratchet is the basis for recombination as necessary for the perpetuation of life.

Multiple allelism: The capacity of a gene to produce more than one variant form of a genetic trait.

Mutating period: A belief by Hugo de Vries that a species experiences nonrandom episodes of mutations, some species-producing and others variety-producing, in unpredictable times.

Mutation: [Muller's definition] A change in the individual gene. [de Vries's definition] A sudden appearance of one or more traits in a species or even the appearance of a new species in a single generation. [In the molecular era] An alteration of one or more bases in a DNA molecule.

Mutator gene: Genes inferred by H.J. Muller to increase or decrease mutation rates. Later analysis in the molecular era suggests that mutations by repair enzymes and infections by transposable elements can lead to such sudden changes in mutation rates.

Muton: The gene as a unit of mutation, identified as a point mutation or site on a fine-structure map.

Negative eugenics: A belief that humans should regulate their heredity by laws, education, and the use of reproductive isolation or sterilization to prevent deleterious heredity from being passed on.

Neutral mutation: A change in the gene that does not alter its phenotype or function.

New synthesis: The phrase used to describe the merging of breeding analysis, cytogenetics, and population genetics with paleontology and comparative anatomy to explain evolution by natural selection.

Nonautonomous mutation: A gene mutation that takes on the normal, not the mutant, expression when its homozygous cells are in a mosaic environment with other cells expressing the normal function.

One gene–one enzyme hypothesis: The finding, in biochemical pathways, that a molecule is assembled in a series of steps, each step involving a specific gene performing a specific enzymatic function, such as adding a side group or forming a ring compound from a linear one.

Operon: A regulatory system consisting of a regulatory protein, a region that recognizes the regulator, and one or more genes that are turned on or turned off by the regulatory protein.

Pangene: A unit of heredity found in a cell that can be transmitted by cell division or by fertilization but that does not move from cell to cell.

Pangenesis: A belief by Darwin that gemmules are produced in each cell and circulated until they are gathered in reproductive cells.

Partial dominance: [William Bateson's definition] A condition of the heterozygous state in which the Aa expression differs from the AA expression, often blending the two traits, as in a red flower crossed to a white flower producing pink flowers in the hybrid. [H.J. Muller's definition] Sometimes special techniques are required to reveal the partial dominance of a recessive trait. A triplo-X or triploid female with $w^+/w/w$ is not as red as one that is $w^+/w^+/w$ or w^+/w^+.

Partial reversion: A partial return to the normal state from a mutant state as in the origin of eosin from a stock of white-eyed flies.

Phenotype: The observable expression of a trait.

Physiological unit: A unit larger than known chemical molecules but smaller than any visible organelles in the cell. Its role is that of forming some component of an organism's heredity.

Plasticity: The tendency of domesticated species to produce numerous variations.

Pleiotropism: The expression of a gene in more than one tissue or organ system.

Point mutation: A mutation that can be localized to a specific site on a genetic map and represented as a point on the line representing the chromosome that bears it. In the molecular era, an alteration of one or more base pairs in a DNA molecule.

Polycistronic system: A complex locus or a pseudoallelic series.

Polygon of instability: The tendency of a hereditary unit to shift from one discontinuous expression to another. Proposed by Francis Galton to explain the origin of both sports and fluctuating variations.

Polymerase chain reaction (PCR): A process using marked primers and a heat-resistant enzyme to make multiple copies of DNA (or genes) as needed. The process is a chemical means of genetic cloning.

Polyploidy: Changes in gene expression associated with extra sets of chromosomes such as triploidy or tetraploidy.

Position effect mutation: An epigenetic mutation in which the gene for a trait is turned on or off by a nearby rearrangement that interferes with its normal regulation. The normal gene can return to function if it is extracted by crossing over.

Positive eugenics: The belief that those with talents, health, longevity, high intelligence, and other socially valued traits should have more children than those who lack those gifts.

Preimplantion genetic screening: The use of a single blastomere in a preimplantation embryo to determine if that embryo is homozygous for a genetic disorder or carries a dominant mutation for a genetic disorder.

Premature doubling of chromatids: An interpretation for the origin of mosaicism by spontaneous or induced mutation. The sperm involved was assumed to be chemically double in its genetic thread but was seen to be a single chromatid when microscopically examined. After 1953, the "doubling" was associated with the DNA double helix.

Prenatal diagnosis: The sampling of amniotic cells, amniotic fluid, or chorionic villi to determine chromosomal or mutational disorders during the first trimester of pregnancy.

Presence and absence hypothesis: The belief, by William Bateson, that a recessive mutation represented the loss or absence of the gene or unit-character for the normal or dominant trait.

Pseudoallelism: The presence of a recombination within members of a multiple allelic series.

Pure line: A population derived from many inbred generations selected for a quantitative trait like size. The resulting pure line can no longer shift the mean of its range.

Recessive mutation: A trait that is not expressed in the hybrid but that can be expressed when rendered homozygous.

Recombinant DNA technology: The use of molecular tools to cut, insert, and splice segments of DNA from a donor source into a target.

Recon: The smallest unit of a fine structure map where no recombination is possible. In practice, adjacent fine-structure sites would provide the smallest map unit dimension within that gene.

Repair enzyme: An enzyme that repairs some specific damage to the DNA. Single-strand breaks, double-strand breaks, cross-linkage of nucleotides, and thymine dimers are removed by different enzymes in the cell.

Repeats: A term used by Calvin Bridges to describe numerous twin sites in salivary chromosomes. Edward Lewis adopted the term as the basis for the origin of pseudoallelic series. Also, it is the physical state of small tandem duplications arising from a primary unequal crossing over, as in the bar eye mutation. In H.J. Muller's analysis, it is the major source of new genes and the later differentiation of duplicated genes in that species.

Residual inheritance: The term used to describe assumed or isolated modifier genes associated with a variable trait identified with its chief gene.

Restriction enzyme: An enzyme that recognizes a specific sequence of DNA and cuts it at that location.

Reverse mutation: A shift from the mutant to the normal state of a gene.

Ribozyme: An RNA molecule that acts as an enzyme.

Rogue: A variation that is unwanted in a breed and is usually culled or removed from breeding with a desired breed.

Semiconservative replication: The mechanism proposed by James Watson and Francis Crick for the replication of DNA in which each strand of DNA serves as a template for the synthesis of a new strand, with each new daughter strand remaining associated with its template strand. This was confirmed by experimental analysis carried out by Matthew Meselson and Franklin Stahl.

Sex chromosomes: The name assigned by E.B. Wilson to chromosomes associated with males or females in Hemipteran bugs. He called these sex chromosomes X and Y, with females being XX and males being XY (or in some species XO).

Sex-limited trait: A trait seen in half the males and none of the females when a cross of the F_1 offspring from a mutant male to a nonmutant female is carried out in fruit flies. Later assigned by Thomas Hunt Morgan to the X chromosome and renamed as **sex-linked trait**.

SNP (single nucleotide polymorphism): A base substitution in a gene used to identify variation in a population or to serve as a marker for nearby genes.

Somatic mutation: The presence of a mutant phenotype in body tissue but not in germinal tissue because of mosaicism that arose shortly after fertilization, producing two lines of cells—one mutant and the other normal.

Split gene: The discovery by Phillip Sharp and Richard Roberts that eukaryotic genes are organized into exons and introns. Exons have the sequences that specify the protein product of the gene. Introns are intervening sequences of DNA that are removed in the processing of messenger RNA (mRNA) from the transcribed RNA of the genes. In most genes there is considerably more intron sequence than exon sequence. The transcribed RNA of the split gene is processed with removal of the intron areas and the assembly of the exon areas. The process is called **splicing**.

Sport: A sudden appearance of a trait that then breeds true.

Sublocus: A region of a complex locus or pseudoallelic series in which the various members can be mapped. The term is an abomination to mathematicians, but arises from the tradition of calling the site of a mapped gene a **locus**.

Swamping effect: A belief that a particular fluctuating variation could not be established by natural selection because the overwhelming majority of members of a population would have opposite effects after mating with it, negating the probability of its survival.

Transcription: The process of copying RNA from DNA by enzymatic means.

Transforming substance: A component of dead fractionated bacterial cells that was capable of restoring the carbohydrate type of the donor and capsule formation to a strain of bacteria that lacked its capsule. The substance turned out to be fragmented DNA. This is also called the **transforming principle**, the terminology used by Oswald Avery, Maclyn McCarty, and Colin MacLeod in their chemical experiment showing this.

Transition mutation: A base substitution of one pyrimidine by another pyrimidine or one purine by another purine.

Translation: The process of aligning amino acids to form a peptide from the nucleotide sequence in mRNA.

Transversion mutation: A substitution of a pyrimidine by a purine or the reverse.

Unfit people: A belief that those with real or alleged hereditary defects are not suitable to reproduce and should be prevented from having offspring.

Unit-character: The term William Bateson used for what later was designated as the gene. Also called **unit factor**.

Variation: Any departure from the wild type or typical appearance of a trait in a species.

X element: An unusual large solitary chromosome first described by Hermann Henking in the fire wasp, *Pyrrhocoris*, and whose function was unknown.

X-linked lethal mutation: A recessive mutation transmitted by a heterozygous female that kills the sons receiving it, resulting in a ratio of 2 females to 1 male.

X-linked trait: A mutation arising on the X chromosome producing a modified 3:1 ratio in which the F_2 mutant class is exclusively male and the nonmutant class has a ratio of 2 females to 1 male.

Index

Page references followed by f denote figures.